An Introduction to Statistical Science in Agriculture

An Introduction to Statistical Science in Agriculture

D. J. Finney
Lecturer in the Design and Analysis of Scientific Experiment University of Oxford

AXIS BOOKS (INDIA)

Published by:
AXIS BOOKS (INDIA)
10-Vyas Ji Ka Nohra, Sardarpura
Chopasani Road, Jodhpur – 342 003
Phone: 91-0291-2642319, Fax: 2643993
E. mail: *axisbooksindia@gmail.com*

AXIS BOOKS (INDIA)

ISBN (... -02-2
Price: ... 25.00

9 789380 604022
Rs. 1095.00 / US$ .00
Published by:
AGROBIOS (INDIA)

Published by: Dr. Updesh Purohit for Axis Books (India), Jodhpur
Lasertypeset at: Yashee Computers and Printers, Jodhpur
Cover Design by: Reena
Printed at: Bharat Printing Press, Jodhpur

PREFACE

For several years I have given an annual course of eight lectures on statistical science to undergraduates reading agriculture at Oxford. In a total of eight hours, detailed instruction in the statistical techniques used in agricultural research is impossible; indeed such instruction might be of little benefit for most of those who attend the course since they will not themselves have need to use the techniques. No student of agriculture to-day, however, can afford to neglect the contribution that statistical science is making to agricultural research. The object of my lectures, and of this book that has developed from them, is to impart a knowledge of the general principles on which statistical science is founded and of the manner in which it enters into so many agricultural problems.

Though I present many numerical examples, the emphasis throughout is on the principles illustrated by the examples rather than on the arithmetic and algebra of calculation. I have long been convinced that a proper understanding of statistical methods as applied in any field of science requires intensive numerical practice, and no doubt a reader who will take the trouble to repeat the arithmetic of my examples will gain considerably thereby. Nevertheless, I am here much less concerned that my readers shall learn to make a t test, or to calculate a standard error of sampling, than that they should appreciate the advantages of careful choice of experimental designs, the chief features of a good sampling investigation, the need for objective statistical analysis of data, the logical basis of significance tests, and related topics. The book is intended to supplement (or even to replace) an elementary course of lectures for agricultural students who are without mathematical training; for many of these, it might represent their complete course of statistics, and for a few it might whet the appetite for fuller knowledge.

The book is designed for consecutive reading and not for occasional reference. Certain sections, marked (†), might be omitted at first reading; the student who finds Chapter 3 difficult or seemingly irrelevant at his first trial might prefer to continue with Chapters 4 and 5, returning to Chapter 3 when he discovers the need for it. Except for a very few simple algebraic formulae, no mathematics appears; although the arithmetical procedures used in the numerical examples are likely to seem strange at first, no reader should have any difficulty in repeating them for himself. Almost all the numerical data included have been deliberately kept simple, in order to encourage the reader to try some analyses for himself. This he can do quite easily by ordinary pen-and-paper arithmetic, though tables of squares, square roots, and logarithms will reduce the time required and a calculating machine will reduce it still further. I have not hesitated to repeat the chief points of important logical arguments in two or more different places, in order to emphasize them; for some this may seem unnecessarily repetitive, but I think that many will be helped to understand the function of statistics by seeing important ideas recurring in slightly different contexts. In all this my chief aim is to show statistical science as a whole — as a single attitude of mind to quantitative problems, and not as a series of isolated methods for particular situations, whose choice must seem arbitrary at first encounter.

I am indebted to Miss M. Callow for checking the numerical examples and drawing the diagrams, to Mrs D. M. Russen for typing and preparing copies of a preliminary version that was distributed to various friends for criticism, to Mr G. L. Hodgson for technical comments and to my father, Mr R. G. S. Finney, for general comments on that earlier version of the book.

D. J. Finney.

Oxford

CONTENTS

[Sections marked † may be omitted in a first reading]

CHAPTER 1

THE NEED FOR STATISTICS

"Though this be madness, yet there is method in't."
Shakespeare: "Hamlet", Act II, Scene 2.

Introduction

In the minds of most people, the word 'statistics' probably conjures up a vision of masses of figures, arranged in an orderly manner page after page; these will be admitted to be of importance for various administrative purposes, but the man in the street finds them neither pleasant nor useful reading. The organization and government of the complex society in which we live require that detailed numerical information should be available on trade and employment, on imports and exports, on the numbers and ages of the population, on accidents, diseases and causes of death. Records of these, and of many other aspects of the social and economic life of the nation, must be kept up to date from year to year, or even from week to week, and must be readily available for those concerned to guide and advise on the well-being of the community. Although the exact amount of planning that ought to be undertaken by the Government is often a matter of heated controversy, no-one will question the need for some measure of planning of national resources. In our thickly populated country, most obvious of all is the need for planning of food supplies, and consequently for reliable information on agricultural resources and production. Herein lies a major reason for the extensive and intensive collection of agricultural statistics, through the medium of the Agricultural Returns made each June, crop estimation, surveys of agricultural practice, and in many other ways.

This book, however, has little or no concern with what are termed *Descriptive Statistics* — namely the extensive tabulation of data collected in order to provide a precise record of particular facts, and eventually to be used as a basis for administrative and executive action. In recent years, a new concept of statistics has arisen, which is most easily distinguished by the name of *Statistical Science*. There is perhaps a tendency to regard statistical science as a difficult and abstruse subject, providing a harmless occupation for professional mathematicians but beyond the understanding of, and having no importance for, anyone else. This view is entirely wrong. Much of the theory of stastistics undoubtedly requires advanced mathematical techniques, but the general logical principles can be appreciated by any intelligent person who is prepared to give some hours to their study and who will accept guidance from mathematicians on purely technical details. Moreover, these principles are not divorced from everyday life or from the other sciences. Rather is it true to say that statistical concepts are an integral part of everyday life, influencing many of the statements we make and the decisions we take, even though we rarely put our arguments into formal statistical language. In most branches of scientific research and practice, where the problems may be more difficult but are at least more exactly specified than in everyday life, the need for logical thought and clarity of expression about all quantitative data is even greater. Day by day, both in professional activities and in private life, we are faced with the necessity of making decisions relating to phenomena that we are unable to discuss in exact terms of cause and effect. To this end we draw upon past experience (both our own and that of other people) in assessing the most likely consequences of certain existing conditions, and then act as in consequence seems best.

Illustrations of Statistical Argument

As an example of the way in which statistical ideas enter almost unconsciously into our everyday action, consider the making of decisions relating to weather. These are often trivial in importance, but, especially in this country, complex in underlying causes. What determines whether a man leaving his home in the morning will or will not carry a raincoat? Certain relevant factors are more or less exactly known to him — the time that he will be away, and the distances that he will have to walk in the open air — but others are uncertain. If there is heavy rain at the time he leaves, no doubt he will take his coat. If not, he will be guided by what he believes to be the likelihood of rain during the important hours. He may rely upon personal estimation, after a glance at the clouds, or he may prefer to use the forecast in his morning paper. However favourable the conditions or the forecast, he can never be *certain* that fine weather at 9 a.m. will not be followed by rain within the next four hours, but, from experience of similar conditions and their consequences in the past, he may judge whether rain is or is not likely. In fact, his decision is probably made in a moment without any conscious logical analysis: in the present state of knowledge, a rapid mental balancing of the many factors involved may be as reliable as a careful examination of meteorological records. The decision, however, is none the less based on statistical considerations, necessarily inexact, of how frequently certain known conditions have in the past been followed by others that would make the coat desirable.

This argument may become clearer by discussion of a closely related problem. Why is July preferred to March as a holiday month? Certainly the hours of daylight will be more in July, but that is not enough. July is usually drier than March, but is not invariably so. Usually July will have more sunshine, but in a particularly unlucky year it might indeed have less than March. In an exceptional year, possibly even the July temperatures would be lower than those in March. Nevertheless, the preference for July continues. It is based on the fact that a decision to take a holiday must be made before the weather is known. Only on August 1st can one say which would have been the better month in that year, but records of former years show clearly that, if July had been chosen consistently in the past, the results would have been more satisfactory than if March had been chosen. If we believe that the general regularity of weather and climatic phenomena in the past will continue, we shall be wise to act in accordance with this experience.

The above illustrations have been deliberately chosen for their general interest, but obviously the same points arise for a farmer deciding on what day to plant his potatoes or to harvest his wheat. As an example of greater agricultural interest, consider the farmer who wishes to choose that potato variety which will give him the most profitable crop on a particular field in the present year. He may not have any exact numerical information, but he will have some knowledge of his own and his neighbours' past experience on the same type of land. He may judge from this experience that a certain variety has usually, but not invariably, been the highest yielder; he cannot be sure of its superiority this year. The previous trials of the variety may have been on various soils, most of which differ a little from that in which he is now interested. He does not know what the weather is going to be, and cannot be sure that conditions will be as favourable to this variety as they were in the past. The variety that seems the best may be more subject to loss from pests or disease than one of its close rivals, and the farmer may make some allowance for his judgment of the probable incidence of these attacks in the present year. Once again the problem is complex, involving these and other factors whose effects cannot be exactly known. The choice must be based upon statistical evaluation of the available evidence, though here also the process is likely to be a rapid integration of all available knowledge by the practical man rather than a formal scientific analysis.

A fourth example may be helpful. Even a simple averaging of values is often essentially a statistical process. However carefully he makes his measurements, a plant physiologist, investigating the growth of a certain species of plant, is not content to base his conclusions on a single plant. Instead, he grows say ten plants under comparable conditions, measures each of them and averages the results. Why should he do this? He is not interested in any one plant for its own sake, but only in so far as it is representative of its species when grown under the conditions that he chooses. He knows that individuals of a species may vary widely in their growth rates, even when the conditions are carefully controlled. Observation and measurement of ten rather than one gives him a better idea of the behaviour of the whole species, for they provide him with a range of variants and reduce the danger that conclusions may be based upon a plant that by chance is very far from typical. The average for the sample of ten plants is used in preference to a measurement for one particular plant, because it is believed to be likely to approach more closely to the appropriate mean value for the whole species (see Chapter 4). Of course, the number ten has no special virtue, and the number actually used will depend upon the time and resources that the physiologist has at his command. Statistical science provides a logical basis for his belief that the mean is a better indicator of the whole species than is a single value, gives quantitative expression to the idea, and enables him to assess the relative merits of different choices for the total number of plants.

Statistical Science

The purpose of statistical science is to provide an objective basis for the analyses of problems in which the data depart from the laws of exact causality. A general logical system of inductive reasoning has been devised, is applicable to data of this kind, and is now widely used in scientific research. Some understanding of its principles (though not necessarily actual skill in its practice) is therefore important both for research workers and for those whose interests lie in the employment of techniological advances resulting from research. Especially is this true of the agricultural and biological sciences. Indeed, much of the modern development in statistical technique has been stimulated by the needs of these fields of research, and in consequence the methods now available are peculiarly well adapted for application there.

The training of an agricultural student or research worker is incomplete unless it gives some attention to statistical science. This short book makes no attempt to teach statistical methods in such a way that the reader can afterwards apply them confidently to his own data. Rather is it intended for the reader who can give only a short time to the subject and wants to come to some appreciation of its meaning. He will be wise to concentrate on the principles, to worry about the details of analyses only in so far as they illustrate these principles, and to accept the purely mathematical technicalities as being the province of specialists. The theory of mathematical statistics has many difficulties, but these are quite distinct from the difficulties encountered by the non-statistician when first he seeks to learn something of statistical science.

This Book and Others

Those who wish to learn enough about statistics to be able to design and analyse their own experiments and other investigations will not find that this book meets their needs. Though numerical examples are given and discussed, these are of the very simplest nature; they deliberately avoid many of the complications that arise in practice, in order to focus attention on basic principles. Nevertheless, I am confident that the reader will come to appreciate the nature of statistical science much more satisfactorily if he repeats for himself each step of the arithmetic of some of the examples.

To attain the higher level of designing and analysing the simpler types of experiment and sampling study is no more difficult, at least to the student who can handle a little elementary algebra, but it does require much more time. Amongst a number of excellent elementary text books may be mentioned: —

C. H. Goulden: Methods of Statistical Analysis (1939: John Wiley & Sons),

G. W. Snedecor: Statistical Methods (1946: Iowa State College Press),

as of particular interest from the agricultural point of view. A little wider in the aspects of biology that it covers, but less detailed in some respects, is

M. H. Quenouille: Introductory Statistics (1950: Butterworth-Springer Ltd.).

Each of these requires prolonged and careful reading, combined with experience, before mastery of the methods is achieved. I want to emphasize most strongly the danger of the 'cookery-book' attitude to statistical science. Attempts to learn just the one or two methods of analysis required in a particular field of research, or to search a text-book for a method applicable to the problem in hand, often lead to completely erroneous treatment of data unless they are supported by a real understanding of the logical principles. The research worker who is prepared to study one of the above books fully, and perhaps to advance further to the classics, amongst which

W. G. Cochran & G. M. Cox: Experimental Designs (1950: John Wiley & Sons),

R. A. Fisher: Statistical Methods for Research Workers (1950: Oliver & Boyd),

R. A. Fisher: The Design of Experiments (1951: Oliver & Boyd),

F. Yates: Sampling Methods for Censuses and Surveys (1949: Charles Griffin & Co.)

are pre-eminent, will add greatly to his powers. If he is not prepared to do so, however, he will be wiser to content himself with a knowledge of principles and to take his problems of technique to the professional statistician. This he should do *before he starts an investigation:* as subsequent chapters will make clear, no statistician can give the best possible service if asked for advice only when an experiment has ended and all the data have been collected.

The reader of this book will come to realize that modern statistical science makes use of many standard tables. Like the tables of logarithms so generally used in arithmetical calculation, these tables present numerical values that have been obtained by complicated mathematical processes. Their practical use, however, may be very simple. Some examples of and extracts from statistical tables appear in later chapters; full details of all that are mentioned can be found in

R. A. Fisher & F. Yates: Statistical Tables for Biological, Agricultural and Medical Research (1948: Oliver & Boyd),

together with brief instructions on their use.

A Warning

This is a book on statistical science, not on agriculture. I have used many numerical examples of the matters I have discussed, but these must be regarded as illustrating principles, not as presenting agricultural facts. Often, indeed, my 'data' are sets of numbers I have invented merely as examples of types of situation and types of calculation. Even where I have drawn upon published records of experiments, I have chosen data to illustrate my argument without thought of whether they represent the fullest or most up-to-date information on the agricultural matters to which they relate. Thus Table 3.3 contains data published as long ago as 1910, and deliberately omits part of the records not relevant to the argument; Fig. 6.3 and the accompanying analyses are for one experiment on the manuring of kale that happened to be convenient, and I am not in the least concerned with whether it showed typical responses to ammonium humate. I have purposely not described fully the agricultural conditions under which experiments were done, as this might unduly have persuaded the reader to regard the conclusions as guidance on agricultural practice.

CHAPTER 2

SOME PROBLEMS OF RATES AND FREQUENCIES

"Although I am almost illiterate mathematically, I grasped very early in life that anyone who can count to ten can count upward indefinitely if he is fool enough to do so."
Robertson Davies: "The Table Talk of Samuel Marchbanks"

Many types of experimental and observational data relate to frequencies of occurrences, that is to say to counts of particular classes of results rather than to measurements of weight, height, length, or other quantitative characters. If ten pennies are spun, the number showing 'tails' must be one of the eleven integers 0, 1, 2, 3, ... 9, 10; no trial can give 7½ tails! So, if ten animals receive a treatment to cure them of disease, the number of recoveries must be an integer. Data of this kind are often expressed as rates or percentages, since a statement that a fraction 0.4, or 40 %, of the animals recovered may be more convenient than the actual number of recoveries for comparison with information on a second treatment in which a different total number of animals was tested. This does not alter the fact that, in a particular trial, only a limited number of alternative results is possible; when 10 animals are treated, the recovery rate observed cannot be 75 %! Although data subject to this limitation may present special difficulties in respect of mathematical theory, at the level of this book they are perhaps the easiest to discuss.

A Contingency Table

As an example, consider an experiment of very simple character. A new treatment is proposed for the protection of animals exposed to the risk of a certain disease. As a test of its efficiency, a group of animals is treated and the number that subsequently die from the disease is recorded. Thus an estimate of the death rate among treated animals is obtained, but further information is necessary before the usefulness of the treatment can be assessed. Unless the disease is known to be invariably fatal for untreated animals, or the treated animals show no deaths in a large group tested, the experimenter will be in doubt as to whether the death rate is any less than a comparable oup of untreated animals would show. It is therefore usual to include in the experiment a second group of animals as 'controls', these receiving no treatment but being otherwise comparable in every way.

Suppose then that the experimenter can use 300 animals, and decides to allot 100 of these to the treatment, the remaining 200 to the control group. After an interval long enough to allow effects of the treatment to appear, he records the number of deaths that have occurred in each group. The results might be as shown in Table 2.1, and further discussion will be based upon these hypothetical data. Such a table is known as a *contingency table,* the form illustrated here being a 2×2 contingency table since it classifies all the animals into one or other of two rows (untreated or treated) and one or other of two columns (surviving or dead).

Table 2.1: Hypothetical results of an experiment on protection against disease.

Group	No. of animals			% dead
	Surviving	Dead	Total	
Untreated	152	48	200	24
Treated	88	12	100	12
Total	240	60	300	20

A first inspection of Table 2.1 gives the impression that the treatment has reduced the mortality rate, since the experiment shows 24 % of untreated animals dead and only 12 % of treated. Is this genuinely a result of the treatment, or is it merely chance? Even if the treatment were entirely without any real effect, to find exactly the same death rates in the two groups would be something of a coincidence, and therefore the mere fact of inequality must not be construed as demonstrating a treatment effect. If only 26 of the untreated group had died, a 13 % death rate, few would attach much weight to the evidence that treatment reduced mortality. The situation is similar to that for an experiment that we might make by spinning two coins. If we spin both 100 times and record the results, we shall not necessarily find that each shows 'tails' exactly 50 times; if the first gives 51 tails and the second 48, no one would regard this as convincing evidence that the first has a bias in favour of tails and the second a bias in favour of heads. The difference in mortality rates shown in Table 2.1 is real enough as a summary of the results of the experiment, but the question of practical importance is whether or not the experimenter can say with confidence that future applications of the treatment, under similar conditions, will give other animals a better chance of survival than if they were not treated.

The answer to this question clearly depends on how large a difference between the two death rates might reasonably be attributed to chance. If the actual difference in Table 2.1 (24 %—12 %) is greater than this, the experimenter will conclude that a real effect of treatment is involved; if not, he must suspend judgement and regard his case as 'not proven'. No simple rule of universal application can be given in the form that a difference that exceeds a particular value (such as 10 % or 20 %) is too great to have occurred by chance, for the number of animals involved is obviously relevant. As an extreme instance, suppose that only three animals were used in the experiment, two untreated and one treated; if one of the untreated was the only animal to die, the two death rates would be 50 % and zero but this apparently large difference might very easily be due to chance*.

One way in which the magnitude of differences that might occur by chance may be studied is by direct trial in a process of experimental sampling. Fortunately this never needs to be used, as it would often be prohibitively laborious, but careful consideration of the idea is valuable for the light it throws on the method finally adopted. Suppose that the difference in Table 2.1 is entirely due to chance, the treatment having had no real effect. This supposition is made in the hope that it may be disproved by a *reductio ad absurdum* argument similar to that often used in elementary geometry. However, instead of demonstrating the falsity of a hypothesis by the impossibility of consequences deduced from it, all that can be done in a statistical problem is to show that the observations under study would be very unlikely to occur if the hypothesis were true and that therefore the hypothesis may reasonably be denied. The argument will be explained in the next three paragraphs, which rest upon the hypothesis of no effect of treatment.

If the treatment is without effect, the two groups in Table 2.1 may be combined to give a single estimate of the death rate; in all, 60

* The novice in statistics should beware of misleading others, or being himself misled, by the use of percentages. Of three fields of winter wheat on a farm, one had to be re-sown the following spring: how much more impressive to say that 33 % of the fields (or even 33⅓ %) had to be resown! Yet how little weight should the statement carry by comparison with a statement that of 300 fields in one part of the country 100 required re-sowing! Expression of information as a percentage is often convenient, but can never of itself increase the value of the information and may be seriously misinterpreted if the total number of cases on which the percentage has been based is forgotten.

out of 300 animals died, or 20 %. Now consider the repetition of experiments of the same kind in a population subject to a 20 % death rate. We may simulate this by taking 300 pieces of card, identical in size and shape, of which 200 are white and 100 red; we mix these thoroughly and withdraw a sample of 60. The two colours correspond to the untreated and treated animals, the sample of 60 to the deaths, and classification of the sample into white and red cards enables a table like Table 2.1 to be formed. We then mix the 300 cards again and draw a new sample of 60. The process may be repeated many times, each trial giving a pair of "death rates" ("untreated" and "treated"); by the nature of the experiment, any difference between the two rates can only be due to chance. There are 61 possible results for a trial, the number of red cards in the sample being any of the numbers 0, 1, 2, 3, . . . 59, 60, but some of these will occur very rarely relative to the others. From our large number of trials, we may compare the frequencies of the various results and so obtain a measure of the extent to which a difference of any specified magnitude may be expected to occur by chance.

For each trial we may record the difference between the "death rates" for white and red cards, and inquire how often this is as large as the value from the experiment in Table 2.1. Since we know that chance alone is responsible for any difference in the rates for the card sampling, we shall thus obtain a measure of the rarity of the data in Table 2.1 if the treatment were without effect. Now the difference in rates for white and red cards will be as large as 12 % only when the number of red cards in the sample of 60 is *either* not greater than 12 *or* not less than 28 (for allowance must be made for the possibility that, by chance or otherwise, the death rate among the treated animals is the higher). As may be proved mathematically, or demonstrated practically by anyone who prefers the great labour of drawing the samples to reliance upon the mathematician, on an average little more than two trials per hundred satisfy this condition. Clearly at least 1,000 trials would be needed before the card-drawing process gave any trustworthy determination of the frequency; the alternative mathematical calculation is much quicker, but it will not be described here. Hence, in the absence of any real effect of treatment, a result as extreme as that recorded in Table 2.1 will be rare, although it is not impossible of occurrence in any one instance. Even though so low a frequency is logically quite different from impossibility, it might often be considered to constitute an improbability so extreme as to warrant rejection of the hypothesis on which it was based. The experimenter would conclude that the treatment had affected the death rate, for the alternative belief was tenable only if a very improbable event had occurred.

Very different would have been the conclusion if Table 2.1 had shown 44 deaths amongst the untreated animals (22 %) and 16 amongst the treated (16 %), a difference of 6 %. On the hypothesis that the treatment had no effect, the same procedure as before would show that results as extreme would occur rather more often than once in four trials. This occurrence is neither an absurd nor even an unusual consequence of truth of the hypothesis, the experimenter could not then reasonably reject the hypothesis, and he could not consider his experiment to have demonstrated convincingly any advantage for the treatment.

Tests of Significance

This kind of argument is known as a *test of significance*. As a first step to demonstrating that a treatment has had some effect, we formulate a *null hypothesis* of the opposite, namely that the treatment had no effect. On the basis of the null hypothesis, we investigate the frequency with which results as extreme as those observed would occur in repetitions of the experiment; here 'results as extreme as those observed' means results showing a difference between the treated and untreated groups as great as or greater than that in the actual experiment. Such an investigation is in practice usually purely mathematical, but we ought always to have in mind the possibility that the answer could be reached directly, by simple but laborious sampling of pieces of card in a scheme like that discussed in the last section, provided that the labour of making a large number of trials could be undertaken. If the frequency of the extreme results is high, the null hypothesis is tenable: the experimental data might easily have arisen even though the treatment had no effect, and therefore the experiment has not given convincing evidence that an effect exists. If the frequency is very small relative to that of less extreme results, the null hypothesis is regarded as being inadequate to explain the experiment and is rejected: the difference between the treated and untreated groups is said to be *statistically significant* or, with an implied special meaning for the word, just *significant*.

This still leaves open the question of how small a relative frequency must be before it is considered to justify rejection of the null hypothesis. Even the most extreme results can occur by chance, so that we can never achieve absolute disproof of the null hypothesis: we reject a null hypothesis only because the experimental results would be very improbable of occurrence if it were true. A common practice, especially in agricultural research, is to use a relative frequency (or *probability;* see Chapter 3) of 1 in 20 (0.05) as the criterion, though some prefer to use 1 in 100 (0.01). These two are sufficiently well established for unqualified use of the word 'significant' to be taken as relating to the 0.05 level, and for 'highly significant' to be taken as relating to the 0.01 level. There is nothing either sacred or mysterious about these levels: they have merely appealed to the body of statisticians and users of statistics as convenient and often appropriate. Any experimenter is at liberty to make his own choice of level, but if he departs from convention he should state his usage explicitly. Much depends upon the use to be made of the experiment, or the action to be taken as a consequence of it. In the experiment reported in Table 2.1, for example, if the treatment under test were some small and inexpensive modification of diet, even slight evidence of a benefit from it might justify recommendation of the treatment as a general practice, or at any rate extension of the experimental programme. Especially is this attitude appropriate in medical research, where a treatment may be known *a priori* to have almost no likelihood of harming the patient and a small indication of benefit is worth following up even at the risk of being a false trail; in these circumstances, possibly a 1 in 10 probability level may deserve to be taken as significant. At the other extreme, if the treatment in Table 2.1 were a radical alteration in the usual system of management of sheep or cattle, no extensive action would be taken unless the advantage for the treatment were demonstrated beyond all reasonable doubt. Even a 1 in 100 probability level for the significance test might then not justify the recommendation that widespread and expensive changes in general practice be made; either a small frequency, such as 1 in 1,000 might be required, or, a rather better policy, further experiments might be taken under slightly different conditions. Moreover the magnitude of the benefit from treatment would also need consideration, since an effect might be statistically significant in a large experiment yet too small to be of practical importance.

At this point, the reader may feel so confused that he doubts the logic of using any level of probability in the manner described. He should return to Table 2.1 and ask himself why he would regard a result that 60 untreated and no treated animals died (30 % and 0 %) as convincing proof that the treatment had an effect. Presumably the only reason is that such a result would be extraordinarily unlikely if the treatment were without effect: the common-sense conclusion is in fact based upon a test of significance! So if 59 untreated and 1 treated animal had died, the difference would again be extraordinary if the null hypothesis were true. No experimenter can avoid basing conclusions on arguments of this kind, and the only problem concerns the choice of the level of probability or relative frequency that is to be regarded as critical: shall he insist on a probability less than 1 in 10,000 and risk failure to classify a difference as significant unless it is very large, or shall be content with a probability less than 1 in 10 and risk occasionally judging as significant a difference that is entirely due to chance? The statistician can give no decision on this, although he can draw the experimenter's attention to what has been found useful in the past.

Application of a significance test may be regarded as consisting of the following four steps:

(i) Formulate the null hypothesis whose contradiction represents the effect that is being sought.
(ii) Choose the level of probability that seems most likely to be useful in respect of the type of decision that depends upon the experiment.
(iii) With the aid of either mathematical theory or trial sampling such as that described in the last section, determine the probability that, if the null hypothesis were true, chance departures from perfect agreement with that hypothesis would be as great as those in the actual experiment.
(iv) If the probability obtained in (iii) is smaller than that chosen in (ii), reject the null hypothesis; if not, suspend judgement.

Note that step (ii) must precede step (iii), if the objectivity of the test is to be preserved! Although step (iii) may depend on difficult mathematical theory, its performance on any particular experimental data usually requires only elementary arithmetic.

A significance test *never* proves that a treatment had no effect. In an experiment like that of Table 2.1, a non-significant difference between the treated and untreated groups would not prove that the true mortality rates were exactly equal; it would merely show that any real difference was fairly small and might even be in the opposite direction from that desired.

The χ^2 Test

Construction of the relative frequencies for the various outcomes of an experiment by drawing samples of cards, as described earlier, is far too laborious for practical use. If it were done for the experiment in Table 2.1, the work would be of no future use for other experiments with different total numbers of animals. Arithmetical calculation of the frequencies from the mathematical theory is also laborious, unless the numbers of animals are small, and some easier approach is desirable.

Table 2.2: Results of the experiment in Table 2.1 to be expected on an average if death rate is 20 % independently of treatment.

Group	No. of animals			% dead
	Surviving	Dead	Total	
Untreated	160	40	200	20
Treated	80	20	100	20
Total	240	60	300	20

Here the mathematician helps. He begins by forming Table 2.2 to show the average numbers of surviving and dead animals to be expected in the two groups, on the null hypothesis that the 20 % death rate applies to both. In this instance, the numbers are all integers, but they are not necessarily so*. The numbers in a table like Table 2.2 are termed *expectations*.

Table 2.3: Excess (+) or deficiency (—) of the numbers of animals in Table 2.1 by comparison with the expectations in Table 2.2.

Group	No. of animals		
	Surviving	Dead	Total
Untreated	—8	+8	0
Treated	+8	—8	0
Total	0	0	0

From Tables 2.1 and 2.2, he forms a new table to show how far the actual numbers differ from the expectations; Table 2.3 contains the differences between corresponding entries in the other two tables. He then calculates a quantity, usually referred to by the name of χ^2 ('chi-squared'), that is defined by the rule

> "Square each 'discrepancy' in Table 2.3, divide by the corresponding 'expectation' in Table 2.2, and add the results."

---

* Even though individual observations must be integers, their average need not be. In any innings, a batsman scores a whole number of runs, but his average for the season may be 24.71 — or even 0.83! In a flock of sheep, each ewe will have a whole number of lambs, but the average number of lambs per ewe may be 1.4.

He does so because mathematical analysis has shown that, when the null hypothesis is assumed, the relative frequency with which the difference between the mortality rates will exceed any specified amount is closely approximated by a value calculated from χ^2 and not dependent upon either the numbers of animals in the two groups or the over-all death rate. This further calculation is complicated, but tables have been computed that enable the answer to be read without any arithmetic by the user. Whatever the death rates and the numbers in the two groups may be, χ^2 summarizes their essential features in such a way as to allow a good approximation to the relative frequency (of the observed discrepancy from expectation being exceeded by chance when the null hypothesis is true) to be obtained directly without any appeal to experimental drawings of cards. For example, if the experiment of drawing from the mixture of two colours of cards were performed as described above, and χ^2 were calculated each time, about half the values of χ^2 would exceed 0.5, about 1 in 10 would exceed 2.7, about 1 in 20 would exceed 3.8, about 1 in 100 would exceed 6.6, and about 1 in 1,000 would exceed 10.8. Since the discrepancies, and therefore also the value of χ^2, increase steadily as the difference between the two percentage mortalities increases, the types of experimental result showing the most extreme deviation from the null hypothesis will be those for which χ^2 is large. Moreover, these values (0.5, 2.7, 3.8, etc., and many others obtainable from published tables) will correspond to the same relative frequencies when χ^2 is calculated by the same rule from sets of observations that are entirely different in total numbers and in death rates (or any other rates that may be the subject of study).

According to the rule stated,

$$\begin{aligned}\chi^2 &= \frac{(-8)^2}{160} + \frac{8^2}{40} + \frac{8^2}{80} + \frac{(-8)^2}{20} \\ &= \frac{64}{160} + \frac{64}{40} + \frac{64}{80} + \frac{64}{20} \\ &= 0.40 + 1.60 + 0.80 + 3.20 \\ &= 6.00.\end{aligned}$$

From the values given in the last paragraph, this has a relative frequency far less than 1 in 20 but a little greater than 1 in 100 (more exactly, 0.0143), thus agreeing well with the frequency of about 2 in 100 obtainable by the laborious method. A piece of arithmetic taking about ten minutes replaces many hours of work; it leads to a significance test almost identical with the other, but having the advantage that all the hard work has been done by the mathematician who determined what values of χ^2 correspond to certain probability levels and placed these on permanent record. More detailed information on the probability levels of χ^2 has been published, but those stated above suffice for most 2×2 contingency tables. The method of calculation of χ^2 can be simplified so as to give exactly the same answer a little more quickly, but that described here has the merit of easy generalization.

The use of this χ^2 and its tabulated probability levels instead of the complete calculation of relative frequencies mentioned earlier involves approximations that are good only when the numbers of animals are large; a working rule frequently quoted is that none of the expectations (i. e. entries in Table 2.2) should be less than 5. The agreement between the two methods can be improved considerably by one very simple device; this is the reduction of each of the discrepancies between observation and expectation (Table 2.3) by ½ before calculating χ^2, so giving for this experiment

$$\begin{aligned}\chi^2 &= \frac{(-7.5)^2}{160} + \frac{(7.5)^2}{40} + \frac{(7.5)^2}{80} + \frac{(-7.5)^2}{20} \\ &= 0.35 + 1.41 + 0.70 + 2.81 \\ &= 5.27.\end{aligned}$$

The relative frequency according to the mathematical theory is now found to be 0.0217 instead of 0.0143, a better agreement with the true value of 0.0210. This modification, known as the *Yates correction* or the *correction for continuity,* is so easily made that it ought to be generally adopted unless all the numbers of animals are very large. The correction is always ½, whatever the number of observations, and it is always applied so as to bring each discrepancy nearer to zero.

No explanation of why the quantity called χ^2 is used can be attempted here. Note that at least it behaves sensibly, in that (i) squaring the discrepancies allows both positives and negatives to contribute to the evidence against the null hypothesis, and (ii) division by the expectations makes some allowance for the fact that a discrepancy of stated magnitude is less striking relative to a large total than to a small.

Random Selection

In all the above discussion, we have made one exceedingly important assumption without explicitly stating it. We have regarded the 300 animals used in the experiment as independent individuals, of like history and under like management in every way except for the one imposed treatment. A common mistake is to take two groups of animals from different sources and to use one as the treated group, the other as the controls. For example, one flock of 100 sheep might be treated, and a second flock of 200 used as controls. If this is done, the flock becomes the experimental unit, instead of the individual sheep. Unless the experimenter is certain (and, if he is honest with himself, he can seldom attain this certainty) that the natural death rates for the two flocks would be identical in the absence of treatment, either of the significance tests described earlier may be entirely misleading because it ignores the possibility of a difference in death rates having nothing to do with the treatment. As an extreme instance, genetic, nutritional, or other factors in the history of the control flock might have predisposed it to heavy mortality from the disease, while the corresponding factors for the treated flock had just the reverse effect; a large benefit could then appear to result from the treatment, when in reality either heredity or history was responsible. A reversal of the inherent differences between the flocks could nullify a real advantage of treatment, or even produce an apparent ill-effect.

The right procedure is first to allocate a number of sheep to the experiment (say 300), then to decide how many of these are to be treated (say 100 — see below), and finally to select those for treatment entirely *at random* from the total. 'At random' does not mean a haphazard selection of a kind that may allow conscious or unconscious subjective judgement by the experimenter, or qualities such as the activities of individual animals, to play any part; it means a strict objective process of drawing lots, so as to give every animal the same chance of being selected for treatment. The animals might be numbered from 1 to 300, and 100 slips of paper drawn from a well mixed set of 300 bearing numbers corresponding to the sheep. Experience has shown that, unless some process like this is carefully followed, there is a real danger of biases entering into the selection and causing animals inherently more (or, in some circumstances, inherently less) likely to benefit from treatment to be assigned to the treated group. For example, instances have been reported of an assistant, instructed to select a group of laboratory mice at random from a cage containing a large number, instead making his selection by what he believed to be a 'haphazard' picking of animals from the cage; the result was that the tamer and less active mice tended to be chosen, and that these were larger than the average of the cage. If the selected mice were treated and the remainder (or a second group subsequently selected) kept as controls, any effect attributed to the treatment might in reality be a consequence of the greater weight of the treated mice.

Randomization is an insurance against inherent unlikeness of the groups, and the experimenter must take responsibility for any departure from it; he alone has power to insist on strict adherence to the principle from the start of his experiment. No statistical treatment of results can compensate for failure to randomize at the right stage, for neither the statistician nor the experimenter can judge the extent to which non-random groups may differ apart from the effects of treatment. In the extreme instance quoted, that of two different flocks of sheep being used for the two groups, no analysis can unravel the treatment effect (if any) from the inherent differences (if any); only repetition with several flocks in each group, so using the flock as the unit, can make the experiment useful. Moreover, in order to avoid the danger of environmental effects operating differentially on the two groups, the animals ought to be either in a single enclosure during the experiment or, if the nature of the treatment prevents this, in separate enclosures for each animal. When the experiment concerns a disease that may be transmitted from animal to animal, the need for keeping individuals separate is obvious if their independence of response is to be maintained. Even when the need is less obvious, however, a simple division into treated and untreated groups for management during the experiment may result in the appearance of a difference in death rates that has no connexion with the treatment. The importance of considerations of this kind must be assessed in relation to the character of a particular experiment. (See also Chapter 6.)

Experimental Design

Not only does the statistician assist in the interpretation of experimental results, but he can also contribute greatly to the efficient planning of experiments. The experiment now under discussion provides a rather trivial illustration, which is nevertheless worthy of comment; much more will need to be said about this subject later.

In the account of the experiment reported in Table 2.1, no comment was made on how the numbers of animals for the untreated and treated groups should be decided. If the experimenter has 300 animals available, how should he divide them? Some may think that, since the treatment is of primary interest, a majority of animals should be treated. Suppose then that 200 animals were treated and only 100 kept as controls; if the mortality rates happened to be exactly as in Table 2.1, the results would be as shown in Table 2.4. The discrepancies would actually be of the same size as before, and, by calculations which the reader should repeat as an exercise (using the Yates correction)

$$\chi^2 = \frac{(-7.5)^2}{84} + \frac{(7.5)^2}{16} + \frac{(7.5)^2}{168} + \frac{(-7.5)^2}{32}$$

$$= 0.67 + 3.52 + 0.33 + 1.76$$

$$= 6.28.$$

This is larger than the earlier value of 5.27, indicating that the same total number of animals has been used so as to give an experiment more sensitive to the difference in mortality rates and therefore more efficient for discovering the existence of a difference. To carry this to the extreme of assigning, say, 290 animals to the treated group would be absurd, however, because the experimental assessment of the mortality among the controls would be so unreliable as to be little use.

Table 2.4: Hypothetical results of an experiment on protection against disease with death rates the same as in Table 2.1.

Group	No. of animals			% dead
	Surviving	Dead	Total	
Untreated	76	24	100	24
Treated	176	24	200	12
Total	252	48	300	16

Perhaps the most natural procedure would be to divide the animals equally, and put 150 in each group. If again the true mortality rates were 24 % and 12 %, the average value of χ^2 from an experiment so arranged would be about 6.5, so that the experiment is a little more sensitive than either of the unsymmetrical arrangements discussed. This, however, is not quite the best that can be done; in Table 2.5, a series of trial calculations for different numbers of treated animals (out of a total of 300) is summarized, and the largest χ^2 is found to occur when about 175 are treated. In practice, equal division of the available animals will usually be the most useful arrangement. The best possible division will be fairly near to this, but its exact position depends upon the true mortality rates and not merely on the experimenter's opinion as to which group of animals is the more interesting; hence the best division cannot be determined without knowledge of the very point the experiment is intended to illuminate.

Two further points may be mentioned. The first and last entries in Table 2.5 show that a very unequal division of the animals would probably have failed to detect the difference in mortality rates, in the sense that the evidence against the null hypothesis would not have been significant. Secondly, by calculations similar to those used for Table 2.5, the effect of changes in the total number of animals can be studied. For example, if only 100 had been available and 50 had been put in each group, the χ^2 would probably have been about 2.2: even though there were a real difference between the two mortality rates of amount about 12 %, an experiment using 100 animals in all would not suffice to demonstrate this, since, if the 0.05 probability level were used, the evidence might easily fail to contradict the null hypothesis of no real difference. Sometimes the best service a statistician can render to an experimenter is to show him that, unless the effect for which he is looking is much greater than he believes likely, no experiment with the available resources has much chance of demonstrating its existence. At least the experimenter can be spared useless effort when told that, if he has only 100 animals at his disposal, he cannot hope to detect the effect of a treatment that reduces mortality from about 24 % to about 12 %.

Table 2.5: Comparison of χ^2 values obtainable by various divisions of 300 animals for an experiment like that in Table 2.1.

No. of animals treated	χ^2 for 12% & 24% mortality
25	1.2
50	2.8
100	5.3
125	6.1
150	6.5
175	6.6
200	6.3
250	4.0
275	2.0

A Larger Contingency Table

The method of analysis, by means of χ^2, used for the 2×2 contingency table can easily be extended to more complex classifications of data. Table 2.6 contains records of 260 fields cropped with barley in 1942. In the preceding winter or spring, an assessment of the wireworm population of each field had been made, on the basis of which the infestation of each field was classified as 'low', 'moderate', 'high', or 'very high'; at harvest, each crop was graded as 'satisfactory' or 'unsatisfactory', and the table shows the relationship between the two classifications. Wireworm attack is only one of many possible causes of poor yields of cereals, and we need not suppose that all the 58 unsatisfactory crops recorded had suffered from wireworm attack. Any evidence that the proportion of satisfactory crops decreased with increasing wireworm infestation, however, would strongly suggest that barley, although known to be more resistant to wireworm damage than wheat or oats, did to some extent suffer loss on this account. At the foot of Table 2.6 are shown the percentages of satisfactory crops, which decline steadily from 86 % in fields of low infestation to 58 % in fields of high infestation.

Table 2.6: Records of barley crops in 1942, in relation to wireworm infestation.

Crop result	Numbers of fields				
	Infestation				
	Low	Moderate	High	Very high	Total
Satisfactory	94	62	31	15	202
Unsatisfactory ..	15	15	17	11	58
Total	109	77	48	26	260
% satisfactory ..	86	81	65	58	77.7

As before, we must make some allowance for the operation of chance. If we classify 260 fields into four groups quite arbitrarily, without reference to wireworms or any other dangers, the percentages of satisfactory crops would not be identical; once again, χ^2 provides a method of assessing the relative frequency with which deviations from equality as great as those in Table 2.6 would then occur. On the null hypothesis that the success of the crop is independent of the wireworm status, a composite estimate of the percentage of satisfactory crops may be taken from the last column, and is 77.7 % (202 out of 260). Using this proportion, we calculate the average numbers of fields expected in the various cells of the table to be as shown in Table 2.7. For example, 77.7 % of 109 is 84.7; on an average, then, of a group of 109 fields 84.7 would be satisfactory and the remaining 24.3 unsatisfactory (as noted earlier, these 'expectations' need not be whole numbers). By subtraction, the discrepancies in Table 2.8 are formed: $94 - 84.7 = 9.3$. Note that the totals of these, by rows or by columns, are all zero (as in Table 2.3). Using exactly the same rule as before,

$$\chi^2 = \frac{(9.3)^2}{84.7} + \frac{(2.2)^2}{59.8} + \frac{(-6.3)^2}{37.3} + \frac{(-5.2)^2}{20.2} + \frac{(-9.3)^2}{24.3} + \frac{(-2.2)^2}{17.2} + \frac{(6.3)^2}{10.7} + \frac{(5.2)^2}{5.8}$$

$$= 1.02 + 0.08 + 1.06 + 1.34 + 3.56 + 0.28 + 3.71 + 4.66$$

$$= 15.7$$

Table 2.7: Average results to be expected in Table 2.6 if 77.7 % of crops were satisfactory irrespective of infestation.

Crop result	Numbers of fields				
	Infestation				
	Low	Medium	High	Very high	Total
Satisfactory	84.7	59.8	37.3	20.2	202
Unsatisfactory ..	24.3	17.2	10.7	5.8	58
Total	109	77	48	26	260
% satisfactory ..	77.7	77.7	77.7	77.7	77.7

Table 2.8: Excess (+) or deficiency (—) of the numbers of crops in Table 2.6, by comparison with the expectations in Table 2.7.

Crop result	Numbers of fields				
	Infestation				
	Low	Medium	High	Very high	Total
Satisfactory	9.3	2.2	—6.3	—5.2	0
Unsatisfactory ..	—9.3	—2.2	6.3	5.2	0
Total	0	0	0	0	0

Although we have calculated a χ^2 by the same rule as before, it now consists of eight contributions from separate observations instead of only four. Nevertheless, it still gives a reasonable measure of the overall deviation of Table 2.6 from Table 2.7. If the number of fields in a particular category differs greatly from expectation (as do the unsatisfactory crops at low infestations in this example), the contribution to χ^2 is correspondingly large. At the same time, division by the expectation makes allowance for the natural and sensible desire to regard a large discrepancy as more important if the expectation is small than if it is large; for example, the unsatisfactory crops at very high infestations make a larger contribution to χ^2 than do those at low infestations (even though the actual discrepancy is smaller), because the discrepancy is much larger relative to the expectation and in calculating χ^2 the expectation is used as a divisor.

Assessment of the statistical significance of data like those in Table 2.6 by reference to an experiment on sampling pieces of card is more complicated, because we should need to use four different colours of card, one for each level of infestation. I shall not attempt to describe how this could be done, and shall be content to use only a χ^2 test. Just as for the 2×2 contingency table, the greater the discrepancies between observed and expected frequencies are, the greater will be the value of χ^2. We may intuitively feel, however, that since χ^2 now contains eight items instead of four, even purely chance variations from expectation could make it larger than previously. This is true, and the test requires a slight modification. Fortunately, study of the mathematical theory of χ^2 shows that the values corresponding to the various levels of relative frequency depend upon the number of contributing classes in a very simple way. Table 2.9 contains the 1 in 10 (0.1), 1 in 20 (0.05), and 1 in 100 (0.01) levels arranged according to the number of *degrees of freedom* (d. f.). The number of degrees of freedom is a number obtained by multiplying together one less than the number of rows in the original table and one less than the number of columns. For an r×c contingency table, that is to say a table with r rows and c columns (exclusive of the row and column for the totals), the number is (r — 1) × (c — 1). Table 2.6 is a 2 × 4 table, 2 grades of crop result and 4 of wireworm infestation, and so has 1 × 3, or 3 degrees of freedom. Table 2.9, which is extracted from much larger and more detailed tables of values that have been calculated from the mathematical theory, states that a χ^2 with 3 d. f. (often written $\chi^2_{[3]}$) will exceed 7.8 on an average once in 20 trials and will exceed 11.3

Table 2.9: Some significance levels for χ^2.

Degrees of freedom	Relative frequency (probability)		
	0.1	0.05	0.01
1	2.7	3.8	6.6
2	4.6	6.0	9.2
3	6.3	7.8	11.3
4	7.8	9.5	13.3
6	10.6	12.6	16.8
8	13.4	15.5	20.1
10	16.0	18.3	23.2
15	22.3	25.0	30.6
20	28.4	31.4	37.6
25	34.4	37.7	44.3
30	40.3	43.8	50.9

once in 100 trials. The value 15.7 is far beyond these, and would be generally regarded as clearly justifying rejection of the null hypothesis. The data in Table 2.6 are not consistent with a belief that the proportion of successful crops is independent of the wireworm infestation.

Final Notes

No attempt is made here to give a full explanation of degrees of freedom, although this matter is vitally important to statistical theory. For the present purpose, the number of degrees of freedom may be regarded as nothing more than a key to the right part of Table 2.9.

The method illustrated for a 2×4 contingency table can be generalized to a table with more rows and columns (e. g. 3×5, 4×4), the number of degrees of freedom being always $(r - 1) \times (c - 1)$. It should *not* be applied when some of the expectations (i. e. values corresponding to Table 2.7) are very small, and special statistical analysis may then be needed.

The adjustment of reducing discrepancies by ½ applies *only* to 2×2 tables, and must never be used with others.

The χ^2 test as illustrated in this chapter must *never* be used for data that are measurements of lengths, weights, times, etc. instead of frequencies (see Chapters 3 and 5). Although χ^2 has many uses concerned with measurement data, they are more complicated and will not be discussed in this book.

CHAPTER 3

PROBABILITY

> "I recollect nothing that passed this day, except Johnson's quickness, who, when Dr. Beattie observed, as something remarkable which had happened to him, that he had chanced to see both No. 1 and No. 1000, of the hackney-coaches, the first and the last; 'Why, sir, (said Johnson,) there is an equal chance for one's seeing those two numbers as any other two.' He was clearly right; yet the seeing of the two extremes, each of which is in some degree more conspicuous than the rest, could not but strike one in a stronger manner than the sight of any other two numbers."
>
> *Boswell:* "The Life of Samuel Johnson".

The Meaning of Probability

In Chapter 2, the word *probability* has been used without any clear explanation of its meaning. Ideas of probability are so important to the statistical analysis of experimental and observational data that we need to examine them carefully before we proceed further. The approach used here may at first seem academic and impractical, but unless we establish a reasonably firm logical basis for our later arguments we shall be led into absurdities.

The word 'probable' is to be found in the vocabulary of most adults, and presumably they use it with some generally recognized meaning. Nevertheless, as soon as an attempt is made to measure probability numerically, difficulties are encountered that have taxed the brains of many mathematical philosophers. We cannot now explore the vast literature of the subject, but we need to recognize two schools of thought, those who regard probability as a 'degree of rational belief' and those who regard it as a 'limit of relative frequency'. Broadly speaking, the first school is concerned with questions in which subjective judgement plays a large part, such as statements relating to unique occurrences. Whether or not the degree of probability in the statements

> "The plays usually attributed to Shakespeare were probably not all written by one man",

and

> "Large scale industrial use of atomic energy will probably not be practicable before 1970"

can be measured on a numerical scale may be debated; even if it can, the value assigned must depend very considerably upon the beliefs and opinions of the person making the statement.

In experimental science, we have little concern with that type of probability. The scientist is primarily concerned with generalizations about classes of objects and classes of phenomena, and for his purposes the definition of probability in terms of relative frequency is most useful. If a statement is made that

> "This cow will probably produce more than 800 gallons of milk in her next lactation",

or that

> "The yield of wheat on this field will probably be increased if the crop is top-dressed with sulphate of ammonia",

and is regarded as referring only to the unique cow or field, the same problem of personal judgement arises. More usually, however, such a statement is intended to refer to one cow or one field only as representative of a class of cows or fields: the second statement, for example, is based upon evidence that most fields of a character similar to that mentioned will give higher yields if they receive sulphate of ammonia, though some may show no change or even give lower yields.

We can best see the nature of the relative frequency concept of probability from one of the simplest examples. Consider the fall of a coin, 'spun' in the ordinary manner by projection into the air from the fingers. In any one trial, the coin may show 'heads' or 'tails'; if a large number of independent trials is made, however, both types of result occur, and, as the number increases, the proportion of tails approaches a steady value known as the *probability* of tails. If the coin is well-balanced and the spinning process fair, this limit of the relative frequency of tails ought to be ½; this, indeed, is why the spin of a coin is so commonly used in sport, in order to decide which of two teams shall have a certain choice. In a short series of trials, great irregularities may occur; for example, five successive spins of a coin might all show tails. Nevertheless, an experiment should convince the reader that the variations in frequencies of tails rapidly diminish as the number of trials increases. A useful exercise for a class would be for each member to spin a penny 100 times, recording the total number of tails after every trial. After only 5 trials, individuals would vary widely in the proportion of trials that showed tails (possibly as widely as from 0.00 to 1.00); after 10 trials, the variation would be less; and the variation in this proportion would steadily decrease until after 100 trials almost everyone would have a proportion between 0.40 and 0.60. Further increase in the total number of trials would concentrate the proportions still more closely around 0.50*.

An Experimental Approach to Probability

Table 3.1 summarizes an experiment conducted by J. E. Kerrich, who span one coin 10,000 times. In the early stages of his experiment, the proportion of tails varied widely, but the variation rapidly diminished and the proportion appeared to be approaching a limit very

Table 3.1: Results of spinning one coin ten thousand times.

No. of spins	No. of tails	Proportion tails	No. of spins	No. of tails	Proportion tails
1	1	1.00	150	79	0.527
2	2	1.00	200	102	0.510
3	3	1.00	300	154	0 513
4	3	0.75	400	201	0.502
5	3	0.60	500	245	0.490
6	3	0.50	600	288	0.480
7	4	0.57	800	387	0.484
8	4	0.50	1,000	498	0.498
9	5	0.56	1,500	744	0.4960
10	6	0.60	2,000	987	0.4935
15	6	0.40	3,000	1,490	0.4967
20	10	0.50	4,000	1,971	0.4928
30	13	0.43	5,000	2,467	0.4934
40	19	0.48	6,000	2,991	0 4985
50	25	0.50	8,000	3,966	0.4958
100	56	0.56	10,000	4,933	0.4933

close to ½. The approach, however, is never perfectly regular, and the results of some trials take the value further away from 0.50. For example, the first 20 trials contained exactly 10 tails, but the next trial must take the proportion away from 0.50 — actually to 0.476. The proportion was again exactly 0.50 after 50 trials, after 250, after 2,700, and no doubt at many other stages, but clearly can never remain there. The important point is that the variation about 0.50 decreases as the number of trials increases. A second experiment with the same coin might show entirely different results when only a small number of trials was summarized, but as the number increased the proportion

* The sceptic might perform this exercise for himself, taking successively the roles of, say, 25 members of a class.

of tails would approach more and more closely to the value in the first experiment.

In practice, the probability of tails is unlikely to be exactly ½, for the coin will not be perfectly balanced. A very large number of trials would be needed before the bias was disclosed, but indefinite extension of the experiment might eventually show that the limiting proportion was, say, 0.489 or 0.504 rather than 0.500. Mathematical theory proves that, if the probability of tails is near to ½, the number of tails in any moderately large number of trials is almost sure to lie within a range on either side of its average value equal to the square root of the number of trials. For example, with a perfectly balanced coin, the average number of tails in a set of 100 trials is 50*, and in any particular set of 100 trials the number is almost sure to lie within a range of 10 (the square root of 100) on either side of 50, that is to say between 40 and 60; similarly in 1,000 trials the average number of tails is 500, and a particular set of 1,000 is almost sure to have between 468 and 532; in 10,000 trials the expectation is 5,000 and the limits 4,900 to 5,100. The reader should notice that, although the range becomes wider on either side of the average as the number of trials is increased, it becomes *proportionately* narrower: in 100 trials between 40 % and 60 % tails, in 1,000 trials between 47 % and 53 % tails, and in 10,000 trials between 49 % and 51 % tails. If the coin were not perfectly balanced, but had a slight bias in favour of heads, so that the probability of tails was perhaps only 0.48, the average number in 100 trials would be 48 and the number observed in a particular experiment of 100 trials would be almost sure to lie between 38 and 58. So in 1,000 trials the expected number of tails would be 480 with limits at 448 and 512, and in 10,000 trials 4,800 with limits at 4,700 and 4,900. Clearly a series of 100, or even 1,000 trials cannot be expected to demonstrate a small bias in the coin: although the obtaining of 450 tails in 1,000 trials would be a very rare result for a perfectly balanced coin, and so would indicate with near certainty that the coin was biased, a result such as 476 tails in 1,000 might very easily be found with an unbiased coin or with one for which the probability was 0.48 or with many other probabilities. A series of 10,000 trials is enough to make the overlap between the type of result shown by an unbiased coin and the type shown by a coin with a probability 0.48 almost negligible: for example, an experimenter who obtained 4,930 tails in 10,000 trials could say with near certainty that

---

* This average number is the *expected number,* or the *expectation,* in the sense in which these words were used in Chapter 2.

any bias in the coin was not so great as to make the probability of tails as low as 0.48.

The above detailed explanation has been given in terms of a probability that was exactly or nearly ½, but any quantity between 0 and 1 is a possible value for a probability. Examples closely similar to the coin-spinning experiment may be based upon the throwing of an ordinary six-faced die, such as is used in many games of chance. If the die is unbiased, the probability that in any one throw it gives the result '5' is ⅙, in the sense that a long series of independent trials will show a proportion approaching more and more closely to ⅙ as the total number increases. A probability little greater than 0 will be assigned to an event that very rarely happens; for example, the probability that in any one year snow falls in Oxford in May might be 0.02 or less. Similarly, an event of a type that is almost certain to happen, such as hatching out at least one cockerel from a batch of ten eggs, will have a probability nearly 1 (in this instance, about 0.999).

The use of coin-spinning and dice-throwing to illustrate elementary ideas of probability is convenient, because they can be understood by all and the statements made may be confirmed by anyone who will take the trouble to repeat the experiments for himself. Their disadvantage might seem to be that they are of no practical importance except to the professional gambler. This is not true, for, in addition to their value as introductory to more complex ideas, the results they give have an exact correspondence with the laws of Mendelian genetics as applying to inheritance in plants and animals. For example, in Shorthorn cattle, there is a probability of ½ that a calf from a roan cow mated to a red bull will be red; if records were kept of the proportion of red calves in a lengthening series of such crosses, genetic theory indicates that the proportion would behave in a manner similar to the proportion of heads in Table 3.1. Again, the probability that a calf from a roan cow mated to a roan bull will be red is $^1/_4$ (the probability of its being roan is $^1/_2$, and white $^1/_4$).

Probability Distributions

By combining results from several coins spun simultaneously, more complex types of occurrence may be studied, and if the probabilities for the individual coins are known those for the possible combinations can be calculated. Indeed, the calculation of probabilities relating to sets of observations (in particular relating to totals and averages) from knowledge of probabilities for single observations is one of the fundamental contributions of the mathematician to statistical science; from the methods and the results, many important inferences about the best way of collecting and handling numerical data can be drawn. As a simple example of the type of argument, suppose that ten coins are spun simultaneously, and the number showing tails is recorded. One of eleven different results must occur: the number of tails must be

Table 3.2: Results of spinning a set of ten coins.

Number of coins showing 'tails'	Number of times occurring	Probability of this result
0	0	0.001
1	1	0.010
2	3	0.044
3	13	0.117
4	24	0.205
5	31	0.246
6	15	0.205
7	9	0.117
8	3	0.044
9	1	0.010
10	0	0.001

0, 1, 2, 3, ... 9, or 10. If each coin is well-balanced, so as to have a probability exactly ½ of itself showing tails, and if the experiment is repeated many times, the average number of tails to be expected is obviously five, but trial would soon demonstrate that the number is not exactly five every time. By spinning (or, more exactly, shaking in a box) one set of ten coins 100 times each, I obtained the results shown in Table 3.2; I had no reason to suppose that these coins were perfectly balanced, but, as shown below, the results accord very well with a belief that they were very nearly so. Mathematical theory, which must be taken on trust by the reader, shows that, if each coin has a probability ½ of showing tails, the probability that none of the set of ten show tails in any particular trial is 0.001, or more exactly 1/1,024. The probability that all ten show tails is exactly the same. The probability that just one of the ten coins shows tails is 10/1,024, or approximately 0.010, and this is also the probability for nine tails. Similarly the probability of two or of eight tails is 45/1,024. The complete set of probabilities is given in Table 3.2. As usual, these probabilities represent limits of relative frequencies: if the experiment of spinning the ten coins 100 times were repeated often, the average numbers of occasions on which the eleven different numbers of tails were recorded would be expected to be about 100 times these probabilities. Inspection of Table 3.2 should convince most readers that the one experiment actually performed agrees very well with these averages; a form of the χ^2 test, slightly modified from that discussed in Chapter 2, can be used to demonstrate that the deviations from the averages are no greater than could easily occur by chance.

Fig. 3.1 is a diagrammatic representation of the experiment by means of a *histogram;* equal sections of the horizontal base line are

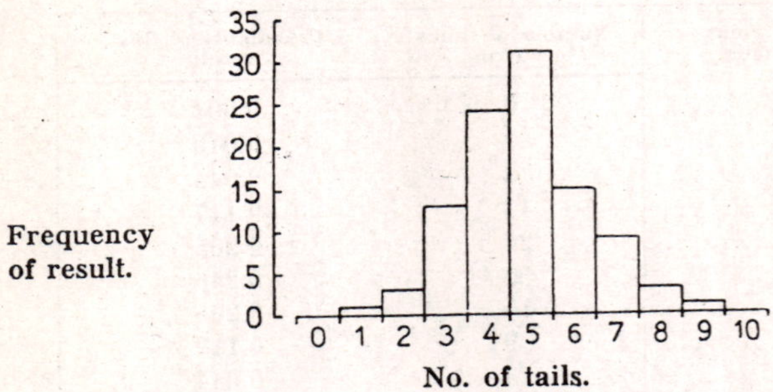

Fig. 3.1: Histogram for coin spinning data in Table 3.2.

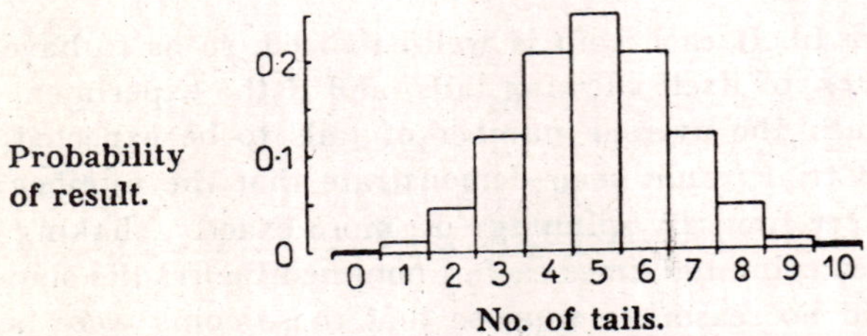

Fig. 3.2: Probability distribution for number of tails in 10 coins, when each coin has probability ½ of showing tails.

made to correspond to the eleven possible results, and on these are drawn rectangles whose areas are proportional to the frequencies with which the results occur. Fig. 3.2 looks very similar; it represents the *probability distribution,* and is constructed from the probabilities in Table 3.2 in the same way as the histogram was constructed from the frequencies. Note that the probability distribution is symmetrical on either side of 5, whereas, on account of the random chances of the experiment, the histogram is not symmetrical.

Results of this type can also be interpreted in terms of genetical observations. Suppose that a red Shorthorn bull is mated with roan cows on ten occasions. As already stated, the probability that any one calf is red is ½, and the probabilities for the ten calves are compounded exactly as are those for the ten coins; the number of red calves will be 0, 1, 2, ..., 9, or 10 with the probabilities shown in Table 3.2 and Fig. 3.2. By collecting records from one hundred such bulls, numbers analogous to those in the coin experiment would be obtained; they would not necessarily be identical with those represented in Fig. 3.1, but they would show similar characteristics and on

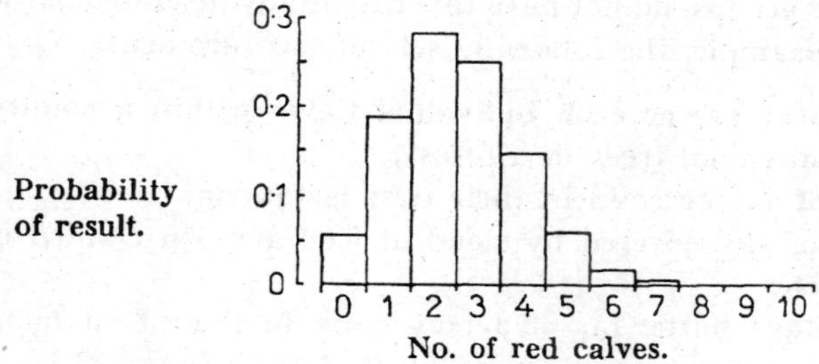

Fig. 3.3: Probability distribution for number of red calves in 10 progeny of a roan bull and a roan cow.

an average the frequencies would agree with the probabilities. If a roan bull is mated with a roan cow, the probability that a single calf is red is ¼; in a set of ten calves from such a cross, there are again the same eleven possible numbers of red calves, but the relative frequencies with which these occur are not the same as before. Fig. 3.3 shows the probabilities for this second situation. Notice that the distribution is no longer symmetrical. Moreover, whereas in the previous example the occurrence of no red calves in a set of ten was very rare, to be expected only about once for every thousand bulls, the probability for no red calves is now much larger and about six bulls per hundred are to be expected to show this. In other words, an observation that none of ten successive calves from roan cows were themselves red would be no more than moderately uncommon if each cow had been mated to a roan bull, but would be so remarkable as to throw doubts on either the correctness of the records or the underlying genetic theory if each had been mated to a red bull.

Discrete and Continuous Variation

The discussion in this chapter has so far been concerned only with observations or experiments for which results are obtained merely by counting of individuals: the data consist of integers. When ten coins are spun, the number of tails can take only one of eleven values, and the result of any one trial can never be 7.3 tails or 0.06 tails! These values can occur as means (averages) of several trials, but not as single observations. A quantity that may vary from one trial to the next but is in this way restricted to a limited set of values is known to statisticians as a *discrete variate*. The restriction usually affects the methods of analysis that are appropriate to the data. Examples of the analysis of data on discrete variation have been discussed in Chapter 2.

Many types of data, however, though capable of arbitrary classification into discrete groups, do not have this essentially discrete character. Consider, for example, the following sets of measurements:

(i) Yields of wheat per acre on individual fields within a county;
(ii) Heights of individual trees in a forest;
(iii) Milk yields of Jersey cows in their first lactations;
(iv) Percentages of sky covered by cloud at 9.00 a. m. in Oxford on days in March;
(v) Mean percentage butter-fat of Jersey cows in their first lactations;
(vi) Increases in arable acreage on individual farms in England between 1939 and 1945.

For each of these, the number of possible values is limited only by the accuracy with which measurements are made. Wheat yields might be expressed to the nearest hundredweight per acre, in which case the records would consist mostly of integers between 0 and 30 (with occasional higher values); if instead they were expressed to the nearest one-tenth hundredweight per acre, ten times as many distinguishable numerical values would be possible. The scale is infinitely divisible, and the data are said to relate to a *continuous variate*. Moreover, in the data for discrete variates discussed earlier there was no choice for the scale of measurement: the frequency, or direct count of the number of observations of a certain type, was the only numerical value to be considered. Wheat yields, on the other hand, can be expressed on an unlimited number of different scales, such as hundredweight per acre, tons per acre, quintals per hectare, ounces per square yard; only convention and convenience determine which shall be used, and measurements on one scale may be converted to any other, so altering completely the numerical values but not affecting the data in any fundamental manner. This might seem not to be true of (iv) and (v) above, but the objection arises only because the word 'percentage' has been used; convention and convenience, but no special sanctity, make expression as "parts per hundred" more usual than as "parts per thousand", "parts per seventy", or any other scale.

It is of interest to note that, of the examples given, data in the first three classes must consist of positive values. Some values may be very small, and each class has some real, but not exactly specifiable, upper limit: in (ii), some 'trees' might be only a few inches in height, and all would be distributed over a range extending perhaps to 50 feet, perhaps to 150 feet, but definitely not to 500 feet. In (iv), the percentage may take any value between 0 and 100, but is obviously restricted by these two limits. The percentage butter-fat in (v) is logically constrained to lie between the same two limits, 0 and 100, but in fact is of course limited to the much narrower range from some small minimum value to, perhaps, 6.0. Example (vi) illustrates the possibility of both positive and negative values, the latter representing a decrease in arable acreage; no exact limits can be specified at either extreme, although even without reference to records one may fairly safely assert that no values outside the range — 500 acres to 2,000 acres would be found.

An Example of Continuous Variation

After this explanation of what is meant by a continuous variate, the implications in respect of probabilities must be considered. Table 3.3 contains data on butter-fat percentages measured for 1,333 Ayrshire cows, these being the higher yielders (over 600 gallons in a lactation) from published records for 2,866 cows. The data could be put into a form like that obtained in the simplest coin-spinning experiment by choosing a dividing mark and counting the number of cows above this level. For example, 234 cows (slightly more than one-sixth of the total) have more than 4 % butter-fat; if the number of records were increased, the proportion in this group would be expected to approach a limiting value, just as did the proportion of tails in Table 3.1 except that the limit is now some unknown value in the neighbourhood of 0.17. This division is entirely arbitrary, and could have been made at any point on the scale. Moreover, to summarize the data in Table 3.2 merely by saying that 234 cows gave more than 4.0 % butter-fat clearly omits much of the information supplied.

Fig. 3.4 illustrates what has been said about the possibility of making successively finer subdivisions of the scale of measurement for a continuous variate. Suppose that the butter-fat percentages had been measured (or recorded) only to an accuracy of 0.5, being classified as 2.5 to 3.0, 3.0 to 3.5, etc.; by additions in Table 3.3, the numbers that would be found in each class may be calculated, and these are represented in a histogram in section (c) of Fig. 3.4. This histogram is constructed on the same principle as that in Fig. 3.1, and looks similar in general form except that there are now five classes

Table 3.3: Butter-fat percentages in 1,333 Ayrshire cows yielding more than six hundred gallons of milk.

Butter-fat %	No. of cows
2.6—2.7	1
2.7—2.8	0
2.8—2.9	3
2.9—3.0	5
3.0—3.1	21
3.1—3.2	31
3.2—3.3	66
3.3—3.4	101
3.4—3.5	116
3.5—3.6	162
3.6—3.7	172
3.7—3.8	169
3.8—3.9	147
3.9—4.0	105
4.0—4.1	89
4.1—4.2	64
4.2—4.3	42
4.3—4.4	19
4.4—4.5	7
4.5—4.6	2
4.6—4.7	8
4.7—4.8	2
4.8—4.9	0
4.9—5.0	1
Total ..	1,333

instead of eleven. The difference lies in the fact that the classes are defined not by the intrinsic character of the material but by the manner in which the data happen to be recorded. Fig. 3.4 contains a series of histograms, corresponding to different subdivisions of the scale. An investigator who was content to record the butter-fat percentage to the nearest 0.8 might merely report the numbers in the classes 2.6—3.4, 3.4—4.2, and 4.2—5.0; this would be crude and not very informative, but perfectly legitimate, and section (a) shows the histogram he would obtain. The other sections illustrate histograms obtainable from the same set of cows by investigators who recorded their butter-fat determinations to greater accuracy, section (g) corresponding to the data in the form of Table 3.3. Note that, from the way in which the histogram are drawn, all have the same total area (for this represents the total number of cows), but the area is distributed into successively finer and more numerous classes.

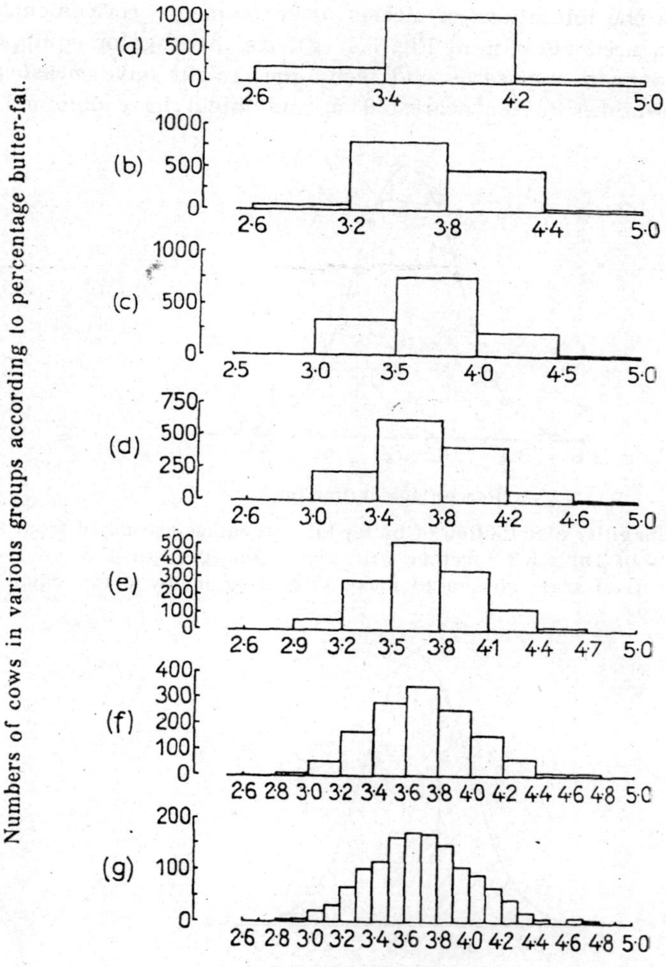

Fig. 3.4: Histograms for butter-fat percentages of 1,333 cows (Table 3.3).

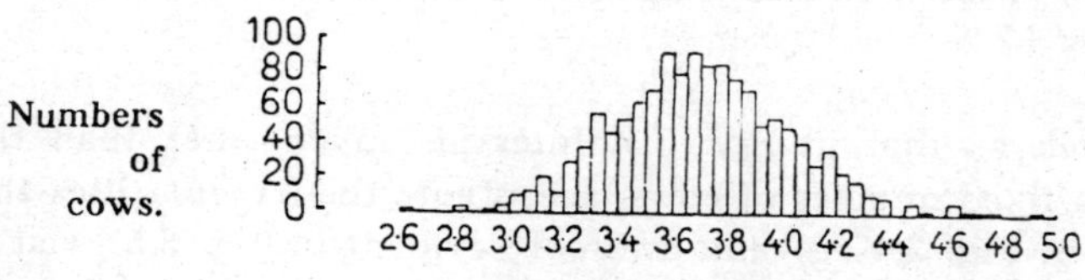

Fig. 3.5: Hypothetical histogram for butter-fat percentages of 1,333 cows, measured to nearest 0.05 %.

Except for the limitations of measuring instruments, refinement of classification need not stop at Fig. 3.4 (g). An investigator equipped with more delicate apparatus and techniques might have measured the fat percentages to the nearest 0.05, and might have obtained a

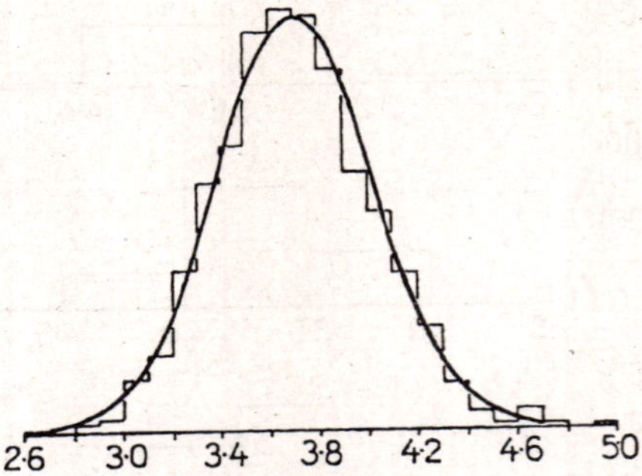

Fig. 3.6: Probability distribution of butter-fat percentage estimated from the data of Table 3.3, together with histogram (g) from Fig. 3.4. [Vertical scale chosen to make total area below curve equal to unity].

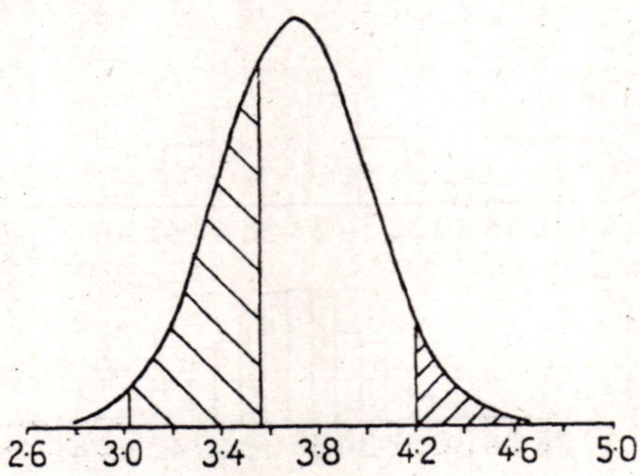

Fig. 3.7: Probability distribution in Fig. 3.6 shaded to show probability of butter-fat between 3.02 % and 3.56 % and probability of butter-fat above 4.2 %.

histogram such as that in Fig. 3.5. Intervals much finer than this would, with a fixed number of cows, accentuate the irregularities that may be seen in Fig. 3.4 (g) and are more evident in Fig. 3.5. Just as a histogram like Fig. 3.1 would be expected to take a more regular form, such as that in Fig. 3.2, when the number of trials is greatly increased, so an increase in the number of cows studied would be expected to make the proportions in the various classes tend to the probabilities for these classes.

If the width of the classes is decreased still further as the number of cows is increased, the outline of the histogram will tend to become indistinguishable from a smooth curve; this curve is the probability distribution for a continuous variate. Fig. 3.6 shows a smooth curve constructed, in a manner that will not be described, to be an approximation to the probability distribution for the Ayrshire cows, and the histogram in Fig. 3.4 (g) has been superposed on it for comparison. Whereas the distributions in Figs. 3.2 and 3.3 could give only the probabilities relating to integral numbers of 'tails' (e. g. the probability of obtaining exactly 8 tails, or of obtaining at least 7 tails, or of obtaining between 2 and 9 tails), since fractional results are impossible and meaningless, that in Fig. 3.6 can give the probability corresponding to any range of values of the fat percentage. The area of the curve between any two vertical lines, expressed as a proportion of the whole area, is the probability that the butter-fat percentage for a particular cow chosen at random will lie between the values on the horizontal scale at the feet of the verticals. Fig. 3.7 shows, by two different types of shading, the probabilities that the percentage lies between 3.02 and 3.56 and that the percentage exceeds 4.20, these being 0.309 and 0.058 respectively.

Enough has been said to give the reader an idea of what is meant by probability and a probability distribution. If he has understood the gist of this discussion, he should be anxious to discover its relevance to practical problems. To describe the inter-relationships of butter-fat percentages in Ayrshire cattle in numerical terms is of little value unless either the fact of having done so or the underlying concept assists the making of inferences from these and other records. This chapter may usefully be re-read at a later stage (say after Chapter 5, and again after Chapter 8), when its fundamental importance will become clearer.

CHAPTER 4

PROPERTIES AND USES OF DISTRIBUTIONS

"... the mathematics [make men] subtile; ..."
Bacon: "Of Studies".

The last chapter may have left the reader thinking that probability distributions could be interesting to the mathematician but are very little help to him. This is far from the truth: although in the simpler routine analysis of scientific data explicit reference to the probability distribution is seldom made, an understanding of the logical processes involved in these analyses requires that the role of the probability distribution be appreciated. If the reader will turn back to Chapter 2, he will realize that what was said there about the analysis of contingency tables might now be phrased more concisely, by relating it to discrete variation as defined in Chapter 3. Similarly, the properties of distributions of continuous variates are the foundation of the statistical techniques needed in the analysis of data on weights, yields, heights, etc. Explanation of these properties is less easy than for discrete variates, since it cannot so readily be put in terms of artificial experiments on repeated sampling from a batch of cards (Chapter 2); it is much simplified by appeal to the differential and integral calculus, an aid that must be eschewed here.

Characteristics of a Probability Distribution

When a diagram of a probability distribution (either discrete or continuous) is inspected, the two most obvious characteristics are the position and the spread. For example, a glance at Fig. 3.6 gives the impression of a distribution round about a value of 3.5 % with a total 'spread' or 'scatter' of about 2.0—2.5 % (measuring the range within which individual values usually lie). These inexact concepts will need to be replaced by more careful definitions, but for a moment the general impression will suffice; someone might prefer to regard 3.4 or 3.8 as a good indicator of the position of this distribution, but no one would argue that the position was 0.2 or 110, or that the total spread was 17.5 or 0.05!

Distributions may differ in position or in spread or in both. Closely related quantities often have distributions almost identical in shape (and therefore in spread) but differing in position. For example, if records of yields per acre were obtained for two varieties of wheat, the probability distributions might be like those in Fig. 4.1 — two identical curves in different positions. On the other hand, if the ton-loads of coal distributed by a somewhat inaccurate coal-merchant were

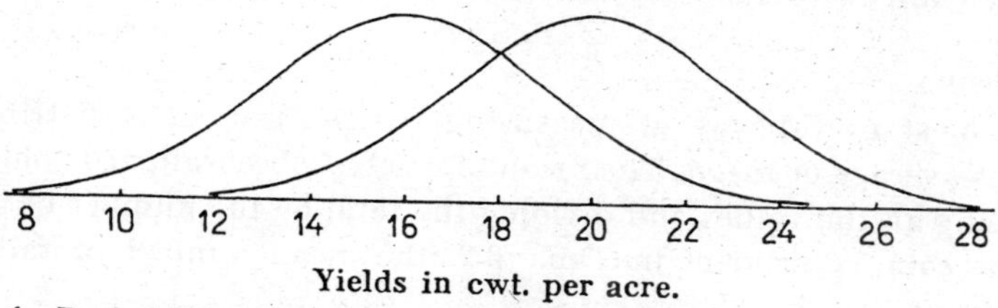

Fig. 4.1: Probability distributions for yields of two varieties of wheat. [Normal distributions with unequal means and equal S.Ds.]

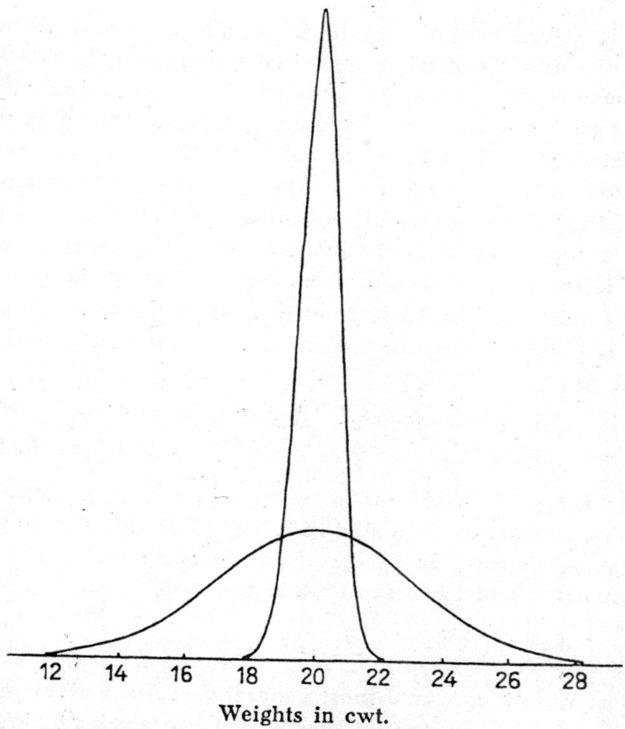

Fig. 4.2: Two probability distributions with means of 20.0. [Normal distributions with equal means and unequal S.Ds.]

weighed more accurately, they might be found to have a distribution identical in position with the better of the varieties from Fig. 4.1 but with the smaller spread illustrated in Fig. 4.2. The 'coal' curve in Fig. 4.2 is basically the same as the 'wheat' curve, having been derived from it by contracting the horizontal scale by a factor of 5 and simultaneously expanding the vertical scale by the same factor; this was done for convenience of illustration only and is not intended to suggest that in reality two such curves would be related in this manner.

The Mean

The most useful way of measuring the position of a distribution is by its average or *mean*. The mean of a set of observations is obtained by adding all the values and dividing the total by the number of items. For the coin experiment in Table 3.2, the mean number of tails per trial is

$$(0\times 0 + 1\times 1 + 3\times 2 + 13\times 3 + 24\times 4 + 31\times 5 + 15\times 6 + 9\times 7 + 3\times 8 + 1\times 9 + 0\times 10) \div 100 = 4.83,$$

each of the eleven possible results being added as many times as it occurs. If the number of trials were vastly increased, so that the relative frequencies of the results approached the probalities, the mean would approach more and more closely to a value known as the mean of the distribution, this being the average of a set of trials whose results have relative frequencies equal to the probabilities. Thus, again for the coin experiment, the mean of the distribution is the same as a mean based upon any set of frequencies directly proportional to the probabilities in Table 3.2; in fact, if the exact probabilities are multiplied by 1,024 a set of frequencies that are all whole numbers is obtained, and the mean of the distribution can then be calculated as

$$(1\times 0 + 10\times 1 + 45\times 2 + 120\times 3 + 210\times 4 + 252\times 5 + 210\times 6 + 120\times 7 + 45\times 8 + 10\times 9 + 1\times 10) \div 1024 = 5.00.$$

Again, the mean of the butter-fat percentages for the 1,333 cows in Table 3.3 may easily be evaluated as 3.704 (if all values between 2.6 % and 2.7 % are called 2.65 %, etc.). The distribution in Fig. 3.6 has been constructed so as to have exactly the same mean.

The Standard Deviation

The most widely used and most generally useful way of measuring the spread of a distribution is by means of its *standard deviation* (S.D.). The older name of 'root mean square deviation' is helpful as indicating the meaning of this quantity. Suppose that the mean of a set of observations is calculated, and then this amount is subtracted from each observation in turn to give a set of *deviations*. If the squares of the deviations are added (each square is, of course, positive) and divided by the number of observations, the result is the mean square deviation*. The square root of this is a quantity, measured in the same units as the original observations, which is obviously in some way indicative of the amount of spread; the standard deviation of the distribution is defined to be the value which the quantity takes when the number of observations is made very large. Some may wonder why all the trouble of squaring and square-rooting is undertaken, instead of the apparently simpler direct calculation of a mean of the deviations themselves: mathematical theory shows the superiority of this measure of spread for distributions of the kind generally encountered.

The distributions in Fig. 4.1 have means 16.0 and 20.0 cwt., the standard deviation of each being 3.0 cwt. Both the distributions in Fig. 4.2 have means 20.0 cwt., the standard deviations being 3.0 and 0.6 cwt. These values do not refer to specific data, but have been arbitrarily chosen as typical of what might be found in the situations described. A standard deviation is always measured in the same units (here cwt.) as the original measurements.

The Greek letter σ is commonly used to represent a standard deviation.

The Normal Distribution

The distributions illustrated in Figs. 4.1 and 4.2 are of a type known to statisticians as *Normal*. The name is unfortunate, and must not be understood as meaning that it is a standard type, deviations from it representing abnormality; in order to emphasize that the word is used in a special sense, it will be given a capital 'N' throughout this book. Nevertheless, many of the quantities commonly encountered in biological research have distributions that are approximately Normal. The alternative name *Gaussian* is sometimes used, in recognition of the extent to which the theory of the distribution was developed by the mathematician Gauss.

For satisfactory definition of exactly what is meant by a Normal distribution, a mathematical equation is necessary, but mere description of the main features of the distribution will suffice here. One of the chief properties of the equation is that it can be so written as to

* In practical calculations, the sum of squares is usually divided by a number one less than the number of observations, for reasons which are irrelevant to the present discussion and are explained in Chapter 5.

depend only upon the mean and the standard deviation: the only difference beween the equations for any two Normal distribution curves is that they have different values of one or both of these quantities. Numerical values that serve in this way to identify a particular distribution curve, as a member of a whole family, are termed *parameters* and are very important in any practical or theoretical discussion of the distribution.

To suppose that all distributions 'normally' encountered are statistically Normal would be a great mistake; every Normal distribution has the 'bell' or 'cocked-hat' shape illustrated in Fig. 4.2, but some distributions of observations are obviously very different from this. For example, the distribution of the percentage of sky obscured by cloud at 9.00 a. m. daily at an English meteorological station might well be something like that shown in Fig. 4.3, extremes of very clear or very cloudy skies being much commoner than intermediate values*. Distributions of similar appearance to the Normal but lacking its symmetry about the mean are often encountered; the distribution of milk yield per cow per week in a dairy herd is likely to be of this kind, with the upper (right-hand) tail' of the distribution much more drawn out than the lower (Fig. 4.4, a *skew* distribution).

* At their first acquaintance with statistical methods, some people have the idea that the distribution in Fig. 4.3 would be Normal if only it were symmetrical, since it could be turned upside down in order to assume something like the characteristic shape! This idea is totally wrong: to turn a distribution curve upside down would completely change its meaning. For example, the curve drawn in Figs. 3.6 and 3.7, which is in fact Normal, tells us that cows with butter-fat percentages near the mean (say between 3.5 % and 3.9 %) are very common relative to those in an equal interval far from the mean (say between 4.2 % and 4.6 %). Fig. 4.3, on the other hand, even if it were symmetrical on either side of 50 % cloud, would tell us that days whose cloudiness was about average (say between 40 % and 60 %) were much less common than days with very clear or with very cloudy skies (say 0 % to 20 %, or 75 % to 95 % cloud). This is a common-sense argument from the general appearances of the two curves and certainly any trick with the *U-shaped* distribution of Fig. 4.3 that falsifies the statements made must be illegitimate. The reader may construct a Normal distribution curve for himself by using a table of logarithms to find values of x and y related by the equation

$$\log y = -A(x-B)^2,$$

where A is any positive number and B is either positive or negative. All Normal curves can be obtained from this equation by inserting suitable values for A and B, and indeed the equation is one way of defining the Normal distribution in mathematical terms. Many equations can be contrived that will give U-shaped curves but all are entirely different from this.

Except where otherwise stated, distributions of continuous variates will hereafter be assumed either Normal or sufficiently nearly so for Normality to be an adequate approximation; for example, the distribution of butter-fat percentages that underlies the data in Table 3.3

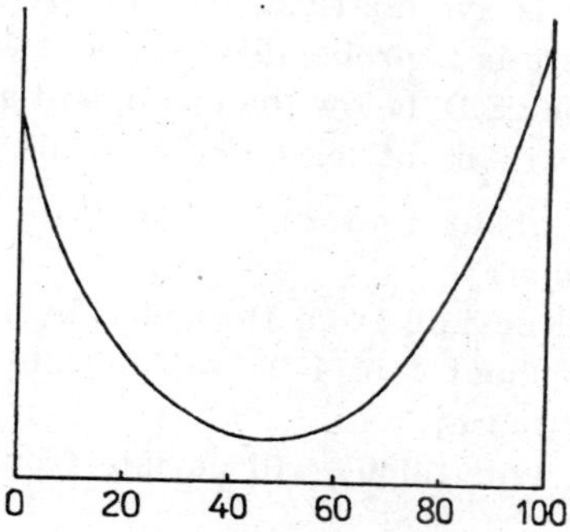

Fig. 4.3: Probability distribution for degree of cloudiness. [U-shaped distribution].

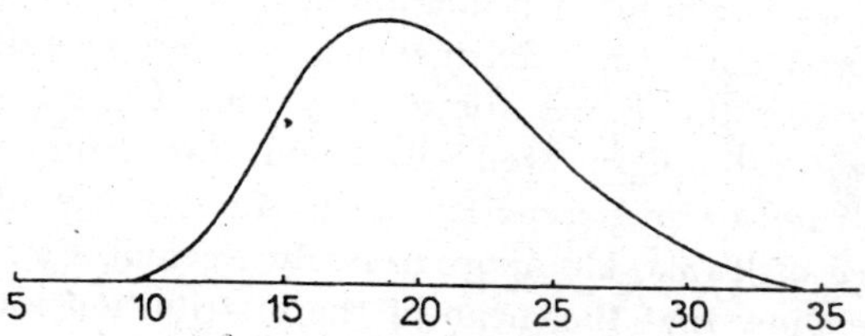

Fig. 4.4: Probability distribution for milk yield per cow per week. [Skew distribution].

cannot be exactly Normal, but these data indicate that a Normal distribution (that drawn in Fig. 3.6) is close to the truth. There are theoretical reasons for this, one of the most important being explained later in this chapter. The mathematical properties of Normal distributions have been more intensively studied than those of any other type.

Normal Deviates and Probabilities

Since a Normal distribution is completely defined by its mean and standard deviation, knowledge of the numerical values of these two parameters enables statements to be made about the relative frequencies of different values of the measurement. Many text-books of statistics contain tables from which information of this kind can be read. Here only a few illustrations will be given. The tables show that:

> 32 % of values (or about 1 in 3) will deviate from the mean by *one* standard deviation or more.

Since the distribution is symmetrical, half of these will lie on either side of the mean: there is a probability of about 1/6 that a measurement will be at least one S.D. below the mean, and an equal probability that a measurement will be at least one S.D. above the mean. Also,

> 4.6 % of values (or about 1 in 20) will deviate from the mean by *two* S.Ds. or more;
>
> 1 % of values will deviate from the mean by 2.58 S.Ds. or more;
>
> 0.27 % of values (or about 1 in 400) will deviate from the mean by *three* S.Ds. or more;
>
> 0.1 % of values (or 1 in 1,000) will deviate from the mean by 3.29 S.Ds. or more;
>
> about 1 value in 15,000 will deviate from the mean by *four* S.Ds. or more.

Clearly deviations in excess of four times the standard deviation are very rare, and will seldom be encountered in short series of records.

In experimental data, these frequencies will not be found exactly: as usual, they are the limits for a very long series. Nevertheless, genuine data will often agree well with them. For example, as already noted, the distribution of percentage butter-fat for Ayrshire cattle is, on the evidence of Table 3.3, approximately Normal. Calculations not described here show that the mean of the distribution must be about 3.70 and the S.D. about 0.32. Hence about 1/6 of the cows would be expected to give percentages below 3.38 and about 1/6 above 4.02; for the 1,333 cows recorded in Table 3, the number below 3.4 is 228 and the number above 4.0 is 234, both of which agree well with 1,333 ÷ 6 or 222. Again, about one cow in 200 would be expected to be below 2.87 and about one in 200 above 4.53 (these differing from the mean by 2.58 times the S.D.); of the 1,333 cows, 4 were below 2.9 and 13 above 4.5, instead of about 7 in each group. According to the theory, an average of one cow per 1,000 should deviate from 3.70 by at least 3.29 standard deviations, or 1.05; of the 1,333 cows in Table 3.3, one was definitely greater than 4.75 and three others lay between 2.6 and 2.7 or between 4.7 and 4.8. The agreement is nowhere perfect, but can be shown to be as satisfactory as would be expected even if the probability distribution were truly Normal; it is naturally poorest for extreme deviations, since the chance occurrence of one very large or very small value may then seem relatively important.

The Distribution of Means

One of the examples of statistical thought used in Chapter 1 related to the use of a mean of measurements made on ten plants, rather than a measurement on a single plant, because the mean would be expected to approach more closely to the mean of the whole species. We are now ready to give a more exact explanation of this statement.

Suppose that, from a large number of individuals having a certain probability distribution for a measurement that may be made on each, we select ten at random and measure them. We might repeat this procedure many times, so as to have many samples of ten, and we could then construct a histogram for the means of the tens. If the number of samples were indefinitely increased, a probability distribution for means of ten would be obtainable. In general, this new distribution will be different in shape and character from the original one. If the form of the original distribution were known, however, an experiment of this kind would enable the distribution of means to be studied, and would show that it had about the same position as the original but a smaller spread; large deviations from the mean of the new distribution will occur more rarely than equal deviations in the original distribution. In other words, the mean of ten will tend to be nearer to the true mean than will a single measurement.

As a student's exercise, a distribution of means could be studied in this way, but as a general practice the method would be intolerably laborious. One of the most valuable types of result in mathematical statistics is that which derives the properties of a new distribution from those of an original distribution used to generate the new. By such an analysis, the distribution of means of ten may be related to the original distribution, without reference to the particular form of the latter. The most important results obtained by the mathematicians are that

(i) The mean of the new distribution (i. e. the 'mean of the means of ten') is identical with the mean of the original,

and

(ii) The S.D. of the new distribution is equal to that of the original divided by the square root of ten (i. e. divided by 3.16).

The first of these results, which is really obvious but for completeness needs to be stated, shows that the mean of ten individuals will on an average have the same value as does the measurement of a single individual, so that the experimenter may legitimately choose whichever procedure is likely to lead him nearer to the truth. The second result indicates the extent of the reduction in the spread of the distribution brought about by the use of means of ten.

These results have been stated for means of ten individuals, but with a slight modification they become more general. If means of N individuals are under consideration, the only change needed is that:

S.D. of means of N = Original S.D. ÷ square root of N.

We customarily denote the individual measurements by the symbol x, and write $\bar{x}$ (read as 'x-bar') for a mean of N of these. The standard deviation of the original distribution is denoted by the Greek letter σ. With this symbolism, the general results may be written:—

(i) Mean of distribution of $\bar{x}$ = Mean of distribution of x,

(ii) S.D. of distribution of $\bar{x} = \sigma \div \sqrt{N}$.

These results apply whatever the form of the original distribution. In general the distribution of means will differ in character from the original; but, if the original distribution is Normal, the distribution of means is also Normal. This most important property of Normal distributions allows us to make statements about the probabilities of specified deviations from the mean of the whole population for any size of sample, provided that we know the S.D. of the distribution of individuals. If the wider curve in Fig. 4.2 represents a distribution of individual values (in this instance, yields per acre on single fields), the distribution of means of 25 fields will have a Normal distribution with standard deviation $1/\sqrt{25}$ (or 1/5) of the original S.D.; thus the narrow curve in the same diagram is identical with this distribution of means. The probabilities for normal deviates that have already been quoted, or the more extensive tables readily available, enable the probability that the mean of 25 fields will exceed any specified amount to be found.

This is a great step forward in the interpretation of numerical data, although a further advance (to be discussed in Chapter 5) is needed before we can apply statistical techniques to the most interesting practical problems. Two examples presented below are somewhat unreal because they assume the investigator to have information of a kind that is rarely available; even so, they are instructive as illustrating the type of argument and drawing attention to what is needed next.

A Simple Significance Test

Suppose that a variety of wheat commonly grown in a certain region is known to have yields that vary from field to field in accordance with a Normal distribution of mean 16.0 cwt. per acre and standard deviation 3.0 cwt. per acre. A new variety is tried on a single field in the same region and yields 20.0 cwt. per acre. Is this adequate evidence that the yielding capacity of the new variety differs from the old, or might the difference found be due merely to chance variations from field to field? In other words, is a single field of the standard variety likely to differ from the general mean by as much as this one field of the new variety? We easily obtain the answer. The new variety exceeded 16.0 cwt. per acre by 4.0 cwt. per acre, or a deviation of 1.3 times the S.D. Now one field in twenty of the standard variety will deviate from 16.0 cwt. per acre by 2 S.Ds. (and so will lie outside the limits 10.0 and 22.0 cwt. per acre), so that a field deviating by only 1.3 times the S.D. would be no surprise; in fact, the probability of a deviation as large as this is 0.18 or about 1 in 6. Thus the one field gives very little evidence that the new variety is a real improvement on the standard.

Now suppose that the new variety is grown on 6 fields, and that the average yield from these is 20 cwt. per acre. If we make the *null hypothesis* (see Chapter 2) that the new variety does not differ from the old in its inherent yielding capacity (i. e. that the probability distributions for the yields of the two are identical), we may assert that the average yield from a set of 6 fields of the new variety ought to have a probability distribution with mean 16.0 cwt. per acre and S.D. $3.0 \div \sqrt{6}$, or 1.22 cwt. per acre. The value 20.0 deviates from the mean by 3.3 times the S.D., and tables show that deviations as large as this have a probability of about 0.001: only one in 1,000 of sets of 6 fields would deviate to this extent. Exactly the same argument as was explained for the test of significance in Chapter 2 applies. If the investigator is willing to take the risk that, when in reality no real difference exists, once in 1,000 times he will claim to have found one, he will regard the probability here determined as significant evidence against the null hypothesis: he will assert that the null hypothesis is untrue, and therefore that the yielding capacity of the new variety does differ from that of the standard. Here the superiority of the new variety is established with very little doubt, and on the evidence most investigators would be prepared to conclude that the result of growing the new variety in place of the standard will generally be an improvement in yield.

The Significance Level

The choice of the level of probability at which significance will be asserted rests upon the same considerations as were put forward in Chapter 2, and these should be reread. For most agricultural research, the 0.05 level is in general use, but, if the result of taking action when in reality no difference exists were especially undesirable, the 0.01 probability level might be used instead. For example, in preliminary experimentation on a large number of new strains of wheat, significance at the 0.05 level (or even at the 0.1 level) might be sufficient justification for subjecting a strain to further trial, the object at this stage being to avoid rejecting anything that might be good; on the other hand, a final decision that a particular new variety was to be widely recommended in place of an old would scarcely be worth while unless the accumulated evidence of superiority was significant at the 0.01 level (except, of course, when the new variety was also considered preferable on other grounds such as hardiness or disease resistance). Many people refer to results at the 0.05 and 0.01 levels as 'significant' and 'highly significant' respectively; though a writer may freely choose some other level if he thinks it more suited to his purpose, these conventions are sufficiently well established for such a choice to need explicit mention in the text.

Summary of Method

The reader should note carefully the sequence of events in this analysis, for it is typical of what is needed in many experiments:

(1) *Formulate the null hypothesis,* in this instance an assertion that, if it were grown on all possible fields, the new variety would give the same distribution of yields as the old.

(2) *Design the experiment:* more will be said about this in Chapters 6 and 7, but here the instruction merely relates to choice of a suitable number of fields for the new variety and random selection of these.

(3) *Conduct the experiment.*

(4) *Summarize the experiment;* here the mean yield of the new variety on a particular set of fields is compared with the known mean for the standard variety, the deviation of the first from the second (according to the null hypothesis) being due to chance variation.

(5) *Determine the S.D.* for the distribution of this deviation, and so assess the probability that so large a deviation would occur by chance.

(6) *Compare this probability with an agreed level* to give a test of significance; if it is smaller, the null hypothesis is rejected.

(7) *Draw conclusions* from the rejection or non-rejection of the null hypothesis.

Significance and Non-significance

The reader must remember that a significance test can *never* prove two varieties to be truly equal in yielding capacity. Had the difference in the experiment described above not been statistically significant, the right conclusion would have been that any real difference in yielding capacity was too small to be clearly detected and might possibly be in the direction opposite to that in the experiment, not that the two varieties had absolutely no difference. A poor experiment, or one that is very small, may fail to demonstrate a variety as better even when it is genuinely superior: in the experiment discussed here, if only two fields had been assigned to the new variety and the average of these had been 20.0 cwt. per acre, the evidence for the superiority of the new variety would not have been significant.

Perhaps the best way of making this clear is to emphasize that significance is a feature of the evidence provided by a particular experiment or set of observations, not an intrinsic property of the new variety or other material under examination. Though the loose form of words "The new variety is significantly better than the old" is often used, this should always be read as meaning "The experiment gives significant evidence that the new variety is better than the old". The novice in statistics is prone to the error of believing that a verdict of not significant' represents proof that the null hypothesis is true; in reality, it merely means absence of sufficient evidence to show anything different. In order to avoid confusion, I advise that the word *non-significant* be used in preference to 'not significant' or 'insignificant', since it makes clearer the distinction from colloquial usage.

Drawbacks to the Experiment

The experiment used as an illustration here would never be satisfactory in this form, although it serves well as a simple example of the fundamental argument. It requires knowledge of the probability distribution for yields of the standard variety that we could not obtain with sufficient precision, since data from an enormous number of fields would be needed. Moreover, weather differences will make the distribution change from year to year, so that to compare yields of a new variety in the present year with those of the standard in a past year could be entirely misleading: if the weather in the present year were unusually favourable for wheat, the new variety would appear at an unfair advantage relative to the standard.

A Better Procedure

For a useful experiment, both varieties must be tested in the same year, under conditions that are as alike as possible. Obviously no determination of the probability distribution for the one variety, a procedure requiring hundreds or thousands of fields, is practicable, and the most that could be done would be to have a small number of fields assigned to each variety. Better than this, because it enables experimental conditions to be kept more nearly constant, is an arrangement that has each variety on several distinct small plots within one field. More precise comparisons of the two can then be made, without the need for large areas of land or for the cooperation of many farmers. In such an experiment, however, the true mean for the standard variety is also unknown: the method of dealing with this difficulty is the next step forward in statistical analysis, but is deferred until Chapter 5.

A Problem in Sampling

A problem distinct from, but closely related to the experiment just described is that of estimating the mean yield of a crop in a certain region. Suppose that experience has shown that, in each of several past years, wheat yields in a region were Normally distributed with S.D. about 5.0 cwt. per acre (for all varieties, now, and not just for one). The mean yields for these past years are irrelevant, and may have differed considerably from year to year provided that the S.D. remained substantially constant. How can this information be used in order to estimate the mean yield of the region in the present year, without requiring yield measurements for every field yet so as to give good assurance that the estimate is within 1½ cwt. per acre of the truth?

The answer depends upon what it means by a 'good assurance'. Unless all fields are measured, the true mean yield must be unknown. Nevertheless, if the investigator is willing to assume that the same S.D. will apply in the present year as in the past, he can state the S.D. for the average of any specified number of fields that he may select as a sample of the total number. He can then deduce the most extreme deviation from the true mean that would occur at a chosen level of probability. By making this probability correspond to his concept of a 'good assurance' (which might, perhaps, be a probability of only 0.05 of being wrong), he can find a measure of his uncertainty of the truth. As a simple example, suppose that yields were measured on 36 fields. The S.D. of the average of these would be $5.0 \div \sqrt{36}$ or 0.83 cwt. per acre. Consequently, the deviation from the true mean that will be exceeded with a probability of 0.05 is 1.96 × 0.83, or 1.6 cwt. per acre*. Hence the average of 36 fields is almost certain to be within 1.6 cwt. per acre of the truth, which is not quite good enough for the requirements of this problem.

By trial with various numbers of fields, the number corresponding to a deviation of 1.5 cwt. per acre could be found. A more direct approach is to use N to represent the required number of fields and to find an equation for N. The S.D. of a mean of N fields is $5.0 \div \sqrt{N}$ cwt. per acre, and the problem requires that the deviate obtained when this is multiplied by 1.96 shall be 1.5: hence

$$1.5 = 1.96 \times \frac{5.0}{\sqrt{N}}$$

or

$$\sqrt{N} = \frac{1.96 \times 5.0}{1.5}$$

or

$$N = \left(\frac{1.96 \times 5.0}{1.5}\right)^2$$

$$= 43.$$

The reader may verify that if more than 43 fields are averaged the deviate for the 0.05 probability level will be less than 1.5 cwt. per acre, whereas for less than 43 fields the deviate will exceed this amount. If the investigator wished to set his standard of 'good assurance' higher, he might require that the probability of his deviation from the mean exceeding 1.5 cwt. per acre should be only 0.01. In the above calculations, 2.58 would replace 1.96, and the number of fields would then have to be 74.

Notice that at any stated level of probability the deviate decreases only in proportion to the square root of the number of fields: in order to halve the deviate, four times the number of fields is necessary.

What happens if the whole region contains less than 43 fields of wheat? Measurement of every field will give the true mean, and no question of standard deviations and errors of estimation arises. How then can an estimate of the precision demanded need more than the total number of fields? The answer is that the method used here assumes the numbers of fields under discussion to be small by comparison with the total number in the region, and an adjustment is necessary when a large proportion of the total is being used. Provided

* A deviation of twice the S.D. was mentioned earlier as corresponding to a probability of 0.046, or roughly 0.05; more exactly the ratio for 0.05 is 1.96.

that the sample is not more than 20 % of the total, the adjustment is not very important, and the effect of ignoring it will only be to overestimate slightly the S.D. of the mean yield of the sample.

Random Selection

In both the experiment and the sampling problem discussed above, fields were selected from all those in the region for special treatment or examination, and were then compared with the whole region. Though not explicitly stated earlier, the validity of the arguments used requires that the fields shall be *randomly* selected from all those available. To explain exactly what is meant by *random selection* is not easy, but the reader must at least distinguish between it and forms of selection in which his own conscious or subconscious judgement plays a part. Ideally, a random selection of, say, 43 fields should be made from a complete list of all fields by some process that gives an equal chance of selection to every possible set of 43; for example, each field may be assigned a number and lots may be drawn for the 'lucky' 43, just as was suggested in the selection of animals for treatment in Chapter 2. Statisticians have prepared 'tables of random numbers' to help such a process.

The dangers of non-random selection should be apparent to anyone who will give some thought to them, and who will refuse to confuse convenience with legitimacy and freedom from bias. For example, to take 43 fields near one town might be very convenient in administration and economical in observation, but could not be regarded as giving a fair cross-section of a much larger region. A new variety tested on such a set of fields might be unduly favoured by a district of high fertility. To take a set of fields because they were offered by farmers who had volunteered to cooperate might similarly be misleading, because their interest in any investigation might arise from a progressive outlook and the practice of superior farming methods. Even to take a 'haphazard' selection by scanning the complete list and occasionally noting one for the sample is fraught with danger; many tests have shown the near-impossibility of eliminating the subconscious judgement of the selector in activities of this kind. Exactly the same difficulties arise in selecting fields or plots for experimental comparison of one variety or treatment with another. The only insurance against biases like these is removal of all subjectivity by recourse to randomization. As will be seen later, some restrictions on randomization are permissible and valuable, but an element of random selection must be present if the statistical techniques under discussion are to be valid.

Except where other methods are explicitly mentioned, random selection is assumed as the basis of all experimental arrangement and sample selection in this book.

†*Non-Normal Distributions*

A rule has been given for deriving the S.D. of a distribution of means from that of the original distribution, this being applicable whatever the form of the original distribution. The theorem that the means of repeated samples of N from a Normal distribution are themselves Normally distributed has been stated. The empirical fact that many of the distributions encountered in biological work are approximately Normal has been noted. What is to be done about the many other distributions that are not Normal?

Though much can be written on this topic, and a great amount of mathematical research has been done, only one more theorem will be stated here. This, however, is exceedingly important in its practical implications: for all but the most exceptional probability distributions, the mean of N values has a distribution that becomes more and more like the Normal in form as N becomes large. Even the mean of three or four may go quite a long way in this direction, and the mean of thirty or forty may be for all practical purposes indistinguishable from Normal. For example, the distribution in Fig. 4.3 is very different from Normal, yet if a new distribution were formed to relate to the mean percentage cloud for a set of several days it would be much nearer to Normality.

In practice, conclusions rarely have to be based on single observations, and the important statistical arguments are those applied to means of several observations or measurements. This theorem helps to justify the basing of inferences on Normal theory, even though the distribution of individuals is not Normal.

Standard Errors

When emphasis rests on the behaviour of a distribution as a whole, the spread is usually described by the standard deviation. In practical problems, the chief interest generally attaches to a particular value in distribution (obtained as a mean of experimental data) and its relation to the mean of that distribution. The standard deviation of the distribution is often called the *standard error* (S.E.) of the particular value; thus, in the wheat experiment, the average yield of six fields of the new variety was 20.0 cwt. per acre with a standard error of 1.22 cwt. per acre. This is written 20.0 $\pm$ 1.22 cwt. per acre ("20.0 plus or minus 1.22").

The distinction between standard deviation and standard error is not of great logical importance, but can be confusing in published work. An author who quotes a numerical value for a standard deviation is likely to be referring to the distribution of individual units (fields, plots, animals, etc.) in his investigation; an author who quotes a standard error is likely to be referring to an average of values for a particular group of units. Wherever any doubt is likely, the careful author will state explicitly: "Standard deviation of individual values", "Standard deviation of distribution of means", "Standard error of mean yield of new variety", etc. Unfortunately for his peace of mind, the reader may sometimes encounter the work of less careful authors; I have seen papers in which most careful reading has left me still unsure whether a value quoted was the S.D. of individual measurements or the S.E. of a mean.

The reader must not be misled by the word 'error' in statistical writings. It does not refer to mistakes, faults, or inaccuracies in the data or the analysis, but to the variability inherent in all measurements of biological material. The standard error of a mean yield is not a flaw that ought to be removed by making a correction; it is a measure of the variability that would be encountered if the process leading to the mean were repeated often, under conditions that were closely similar but inevitably not identical with those for the mean quoted.

Probable Error

In older statistical writings, variation and spread were often measured by the *probable error* (P.E.), a quantity that, for a Normal distribution, is about ⅔ of the S.D. This also was symbolized by the '±' sign. Few statisticians to-day use the P.E., but it may still be found occasionally in current publications, where it is inevitably a source of confusion unless clearly described.

CHAPTER 5

AN EXPERIMENT TO COMPARE TWO VARIETIES

> "This is called the inductive method. Hypothesis, my dear young friend, establishes itself by a cumulative process; or, to use popular language, if you make the same guess often enough it ceases to be a guess and becomes a Scientific Fact."
>
> *C. S. Lewis:* "The Pilgrim's Regress".

The Design of a Simple Experiment

At last we can take the step to an experiment of a type that would be both practicable and practically useful. For discussion here, the experiment is purposely kept very simple; its efficiency could be improved by devices that will be explained later (Chapter 6). Nevertheless, it is one that *could* be performed exactly as described.

In the experimental comparison of two wheat varieties that was discussed in Chapter 4, we had to make the unrealistic assumption that the true mean yield and standard deviation for the standard variety were known. As was later pointed out, this would never be so in practice, and both quantities would have to be estimated from a sample of fields directly comparable with those assigned to the new variety. If whole fields are used as experimental units, however, the wide variability in inherent fertility and other characteristics of a field would tend to obscure real differences between varieties — in other words, would make the S.D. large. By putting the whole experiment on a single field, some parts being assigned to one variety and some to the other, not only can we eliminate much of this variability but also we can simplify the organization of the experiment.

Suppose, then, that within a single field we mark out a number of small plots. Such plots are usually adjacent rectangular areas of identical shape; they may be approximately square, or they may be long and narrow, the governing factors being convenience and experience of the standard deviations obtained with plots of various sizes and shapes (see Chapter 6). Fig. 5.1 represents a piece of land divided into ten plots. Now suppose that several of these plots are selected at random and assigned to the standard variety, A, and the others are assigned to the new variety, B. For reasons similar to those discussed for an entirely different type of experiment in Chapter 2, the best results with a specified number of plots will be obtained by giving half of them to A and half to B; departure from this symmetry enables the yield for one variety to be estimated more precisely, but the gain for the experiment is outweighed by the loss in precision on the other variety. For example, the plots might be numbered, in any order, from 1 to 10 and lots drawn for five of these to have A; Fig. 5.1 shows the effect if plots 1, 2, 4, 6, 7 are drawn for A.

1 A	2 A	3 B
4 A	5 B	6 A
7 A	8 B	
9 B	10 B	

Fig. 5.1: Completely randomized design for the comparison of two varieties.

The crops are then grown, and the ten plots harvested separately. Plants near to the edge of a plot may be affected by conditions on the neighbouring plot. For example, if A grows taller or has a more extensive root system than B, plants near three of the edges of plot 5 may suffer from the competition of neighbouring A plants and may give yields that are not typical of normal agricultural practice in which a large area (several acres, at least) would grow nothing but variety B; at the same time, plants of A adjacent to this plot will benefit from their neighbours' weaker growth. A strip around the edge of each plot is therefore discarded, and only the yield of the remaining central area is recorded. The two varieties are then compared in terms of these yields.

Results of the Experiment

Table 5.1 contains the yields for the experiment described in the last section, plots 1, 2, 4, 6, 7 being shown in the column for variety A and plots 3, 5, 8, 9, 10 in the column for B. Since the plots are equal in size, comparisons could most easily be made in terms of the actual weights of grain harvested from each; the alternative of first expressing each plot value as a yield per acre has been adopted here, in order to present to the reader figures whose magnitudes are more readily comprehended.

Tables 5.2 and 5.3 contain other sets of yields that might have been obtained in a similar experiment. They are shown here for contrast with Table 5.1, so as to assist a discussion of how the nature of the data affects the conclusions that may be drawn. The intention is to consider each table in turn as though it represented the results of the experiment and contained all the information available to the investigator*.

Table 5.1: Yields (cwt. per acre) in first experiment on wheat varieties.

	A	B
	19	23
	14	19
	15	19
	17	21
	20	18
Total	85	100
Mean	17.0	20.0

Table 5.2: Yields (cwt. per acre) in second experiment on wheat varieties.

	A	B
	17	21
	16	20
	17	21
	18	19
	17	19
Total	85	100
Mean	17.0	20.0

Table 5.3: Yields (cwt. per acre) in third experiment on wheat varieties.

	A	B
	9	29
	32	9
	11	14
	12	31
	21	17
Total	85	100
Mean	17.0	20.0

* The data in these three tables are artificial. The numerical values have been chosen so as to make the two mean yields the same in each table, in order to emphasize certain comparisons and contrasts between the three hypothetical experiments.

Suppose that the experimenter obtains the data in Table 5.1; what can he infer about the relative merits of the two varieties? He will no doubt begin by finding the average yields of the two sets of five plots, 17.0 cwt. per acre for A and 20.0 cwt. per acre for B. This suggests the superiority of B, but full inspection of the table discloses some doubt. Two of the plots of variety A gave better yields than the worst plot of B, and were as good as the next two plots of B. In fact, the range of yields for A (14—20 cwt. per acre) has an appreciable overlap with that for B (18—23 cwt. per acre). The experimenter is presumably not interested merely in what happened on these ten plots but in the strength of the evidence that, in the future, fields sown to B will generally yield better than if they were sown to A. No reliable answer to his inquiry can be given merely from inspection of Table 5.1: some might argue that all the B yields are better than the average for A and that this trend would continue, while others, impressed by the overlapping of the better A yields with the poorer B yields, would not consider the evidence adequate for any positive assertion of superiority. Clearly an objective evaluation of the evidence is needed, and this will be discussed after a brief consideration of Tables 5.2 and 5.3.

Suppose now that a second experimenter, working in complete ignorance of the first, performs a similar experiment and obtains the data in Table 5.2. He also will note that the average yields are 17.0 and 20.0 cwt. per acre, an advantage of 3.0 cwt. per acre for B. Moreover, the best of his A yields is less than the worst of his B yields, so that the two series show no overlap; with as many as five plots of each variety, this is a strong indication that the difference is greater than would occur by chance. Each column of yields is remarkably regular, and inspection will suffice to convince him that he may reasonably expect the superiority of B to persist.

Thirdly, suppose that yet another experimenter has the data in Table 5.3. Again the average yields are 17.0 and 20.0 cwt per acre, but now the best of the A yields is better than any for B and the worst of the B yields is as low as the worst for A. In fact, the overlap of the two series is complete, A ranging from 9 to 32 cwt. per acre and B from 9 to 31 cwt. per acre. No critical experimenter would be prepared to make an assertion about a general difference in yielding capacity with any confidence that future trials would confirm it.

Data similar to those of Table 5.2 are what every experimenter hopes for, because from them he can draw clear conclusions without appeal to any laborious statistical analysis. Data similar to those of Table 5.3 are what every experimenter rightly fears: with them, he can again dispense with statistical analysis, but now because no method of analysing the data can possibly allow him to draw a simple yet reliable inference. Careful and well planned experimentation (see Chapter 6) will assist in pushing results that would otherwise be like Table 5.3 in the direction of Table 5.2, but any reader who has experience of experimental work will agree that Table 5.1 is much more nearly typical of how data usually appear. For Table 5.2 and 5.3, such conclusions as can legitimately be drawn are obvious without further analysis, and no statistician would contend that additional calculations are essential. For Table 5.1, however, statistical analysis provides the only way of assessing objectively the weight of evidence in the data, and of reaching conclusions of practical value that are independent of the personal opinion of the experimenter.

The Standard Error of a Difference

Whatever the shortcomings of the difference in mean yields as an indicator of the evidence for a real varietal difference, it is nevertheless, from the single experiment, the best measure we have of the magnitude of any real difference. What we wish to know is whether repetition of the experiment, under similar conditions, will consistently give differences of the same size (or at any rate in the same direction) or whether it will often give differences in favour of A instead of in favour of B; the first state of affairs would indicate a real superiority of B, whereas the second would suggest that the difference observed was primarily determined by chance variations from one experiment to another. We shall in fact formulate the null hypothesis that A and B do not differ in yielding capacity, so that the average difference between them in a long series of experiments ought to be zero. If we knew the probability distribution of differences in means of the two varieties for repeated experimentation, we could examine whether *its* mean* conformed to the null hypothesis by being zero. In practice, we require to make decisions from the evidence of one experiment, and we must therefore submit our null hypothesis to a test of significance

* Think carefully in order to avoid confusion over the several means and distributions involved! For one variety, the five plots are values from a distribution whose mean is the true mean yield of the variety. If we average the yields of the five plots, we have a single value from a distribution of 'means of five' for the variety, a type of distribution discussed in Chapter 4. We are now considering a difference between a mean of five plots of A and a mean of five plots of B, which, when we conceive of repetitions of the experiment, is a single value from a distribution of 'differences of means of five'; that distribution in turn has its own true mean which, according to the null hypothesis, is zero.

by inquiring whether the observed difference in mean yields for that experiment could reasonably be supposed itself a value from a distribution of mean zero or whether it departs from zero too greatly for this hypothesis to be reasonable.

If a mean yield of variety A had been determined from a large number of plots, σ, the S.D. of the individual plots, could be found. The mean yield of the five plots of B would then be assigned a S.E. of $\sigma/\sqrt{5}$, and a test of significance could be applied to the difference between A and B exactly as for the experiment in Chapter 4 where fields were used in place of plots. Here, however, the true mean yield of A is also unknown, and only the information from five plots is available. This has two consequences:—

(i) The S.E. used for testing the significance of the difference in yield must be larger than $\sigma/\sqrt{5}$, in order to allow for the uncertainty in the true mean yield for A as well as that for B;

(ii) The S.D. of the yields of A cannot be known exactly, since only five separate values are known for the yield instead of the whole distribution as was previously assumed.

Suppose that the S.Ds. in the probability distributions of individual plots of the two varieties are σ_A and σ_B. The better varieties of a major economic crop are not likely to be very different in respect of variability, so that σ_A and σ_B may be expected to be of about the same magnitude, but equality is not necessary to the present argument and will not be assumed here. In Chapter 4 was explained how a probability distribution of means can be generated from a probability distribution of individual values. The same kind of process can be used to generate a probability distribution of differences of means. If a random sample of five plots of A and a second random sample of five plots of B were taken from the whole distributions, their means calculated, and the difference (B—A) formed, this difference would summarize the result of one experiment of the kind under discussion. If the process of selecting two samples of plots and forming the difference in means could be repeated a large number of times, a probability distribution for the difference would result. Whatever the nature of the original distributions, this new distribution has the following characteristics:—

(i) Its mean is the difference between the means of the original distributions for A and B;

(ii) Its S.D. is $\sqrt{\left(\frac{\sigma_A{}^2}{5} + \frac{\sigma_B{}^2}{5}\right)}$; more generally, if N_A plots of A and

N_B plots of B were used, the S.D. of the difference would be $\sqrt{\left(\frac{\sigma_A^2}{N_A} + \frac{\sigma_B^2}{N_B}\right)}$

The first of these is almost obvious; the second could be demonstrated convincingly by laborious experiments in artificial sampling, but is easily proved by elementary mathematical theory. As for the distribution of a single mean, the Normal distribution occupies a position of special importance: if the two original distributions are Normal, so is the distribution of the difference of means. If the original distributions are not Normal, the distribution of the difference of means will, under very wide conditions, approximate increasingly well to Normality as N_A and N_B increase.

Variance

The square of a standard deviation or of a standard error is known as a *variance*. The term is useful because variances of independent quantities can be directly added. The two important results now available can be stated as:—

(i) The variance of a mean of N observations is the variance of a single observation divided by N;

(ii) The variance of a difference between two independent means is the sum of their variances.

An Unusable Test of Significance

Suppose that the distribution of plot yields for variety A in Table 5.1 were known to have a standard deviation of 2.3 cwt. per acre, even though the mean of this distribution was unknown. Such knowledge is never available in practice for this type of problem, but brief consideration of its consequences is interesting as a prelude to the final analysis.

The null hypothesis to be made is that the distribution of plot yields for variety B is the same as that for variety A. The consequences of this hypothesis are that, for the distribution of the difference between two means of five plots,

(i) The mean is zero;

(ii) The S.D. is $\sqrt{\frac{2.3^2}{5} + \frac{2.3^2}{5}}$

$$= \sqrt{\frac{10.58}{5}}$$

$$= 1.45 \text{ cwt. per acre.}$$

The second statement follows because, by the hypothesis, the S.D. of the distribution of plots of B is the same as that of A. Hence the experiment may be regarded as giving a single value of 3.0 (the difference between A and B) from a distribution that has mean zero and S.D. 1.45. The deviation from the mean is 2.07 times the S.D., which exceeds the 0.05 probability level (1.96); the probability of so large a deviation occurring by chance (in a Normal distribution, of course) is about 0.038. If the convention of this book is adopted, the difference in yield is said to be too large to be a reasonably probable observation, and therefore to constitute significant evidence for the rejection of the null hypothesis. The experimenter concludes that, under conditions comparable with those of the experiment, variety B is a better yielding variety than A: he estimates the average advantage for B to be 3.0 cwt. per acre with S.E. 1.45 cwt. per acre.

The t Test

When the true S.D. per plot is unknown, as is almost always the case in a real experiment, the data themselves must provide an estimate of it. Chapter 4 mentioned that the S.D. of a distribution was also sometimes called the root mean square deviation, a name that gives the clue to the method of calculation. The left-hand half of Table 5.4 shows the deviations of the individual plot values for A in Table 5.1 from 17.0, the mean of the five, and also the squares of these deviations. Because of the rule by which they are formed, the total of the deviations column must be zero. The squares are all positive and their total depends upon the spread of the individual values about the mean. (At this point, the reader should make the same calculations for variety A in Tables 5.2 and 5.3, and confirm that the sum of the squares is much smaller than 26 for Table 5.2, much larger for Table 5.3). The total is not in itself a satisfactory measure of the spread, however, because obviously it depends upon the number of plots: if

Table 5.4: Deviations and squares of deviations for Table 5.1.

	Variety A		Variety B	
	Deviations	Squares	Deviations	Squares
	2	4	3	9
	—3	9	—1	1
	—2	4	—1	1
	0	0	1	1
	3	9	—2	4
Total	0	26	0	16

the experiment had had ten plots of each variety, there would have been twice as many contributions to the total, which would in consequence have been about twice as large. The obvious step is to form an average of the squares, by dividing by the number of plots. This has one small flaw, in that five plots do not give five *independent* pieces of information on the spread of the distribution, since deviations are measured from the mean of the five*. Full explanation would need algebra beyond the scope of this book, but the appropriate modification is very simple. The sum of the squared deviations ought in fact to be divided by a number one less than the number of plots, this divisor being known as the *degrees of freedom* (it is mathematically very closely related to the concept of degrees of freedom for χ^2). Hence, for the five plots of variety A the divisor is 4. For any number of plots, the rule leads to an unbiased estimate of σ^2, the variance per plot, the square root of which is an estimate of the S.D. If the number of plots were very large, the difference between dividing by that number and dividing by the degrees of freedom would be trivial (for example, to divide by 999 is almost the same as to divide by 1,000); indeed, the rule is equivalent to the way in which the true S.D. would be calculated as a root mean square deviation if the whole distribution were known. The chief practical use, however, is for fairly small numbers of plots, and adoption of the right divisor is then important.

The letter s is used for the estimate of the S.D., corresponding to the true value σ. Hence for variety A

$$\begin{aligned} s^2 &= 26 \div 4 \\ &= 6.5. \end{aligned}$$

In the investigation of whether or not variety B is superior to A, the appropriate null hypothesis (which the experimenter propounds in the hope that his data will controvert it) is again that the two varieties have the same probability distribution of yields. Since the next stages of the argument will be based upon this hypothesis, the data on variety B also give information on the standard deviation. By hypothesis, both columns of Table 5.1 relate to the same distribution, and therefore the

* As an extreme, note that, if the experiment contained only one plot of A, this could give no information on spread, yet the sum of squares (now 0) divided by the number of plots (1) would suggest 0 for the measure of spread. The rule that follows states that the divisor should be 0, the number of degrees of freedom; the division has no determinate result and thus confirms the obvious truth that no information on spread is available.

sum of squares of deviations for B may also be used in the estimation of σ^2. The right-hand half of Table 5.4 shows the construction of a sum of squares, which, taken alone, would give for B

$$\begin{aligned} s^2 &= 16 \div 4 \\ &= 4.0. \end{aligned}$$

Now the two independent estimates of σ^2 need to be combined. Here they could legitimately be averaged to give 5.25. Had the numbers of plots for the two varieties been different, however, as they may be in some experiments, we might reasonably expect that the variety with the larger number of plots would give the larger part of the information on σ^2. This is taken into account by pooling the sums of squares and dividing by the total number of d. f., an operation which here leads to the estimate

$$\begin{aligned} s^2 &= \frac{26 + 16}{4 + 4} \\ &= \frac{42}{8} \\ &= 5.25, \end{aligned}$$

the same as direct averaging would give. The general formula is

$$s^2 = \frac{S\{(x_A - \bar{x}_A)^2\} + S\{(x_B - \bar{x}_B)^2\}}{(N_A - 1) \quad + \quad (N_B - 1)},$$

where N_A is the number of plots of A, x_A is a typical one of the yields of these plots, $\bar{x}_A$ is the average of the N_A yields, $S\{\ \}$ is a shorthand symbol instructing the reader to take the sum of the values enclosed, and quantities relating to B are similarly indicated. This rule is the one to be generally adopted for the best estimate of the variance per plot.

The estimated standard deviation per plot is the square root of s^2, or 2.29 cwt. per acre. (The value assumed for the true S.D. in the last section was deliberately chosen to agree with this). The variance estimated for the mean yield of the five plots of either variety is $s^2/5$, and the rule of variance addition gives the estimated variance of the difference in mean yields as

$$\frac{5.25}{5} + \frac{5.25}{5} = 2.10 = (1.45)^2;$$

therefore 1.45 cwt. per acre is the estimated S.E. of the difference. On the null hypothesis, the observed difference, 3.0 cwt. per acre, is an observation from a Normal distribution with mean zero and S.D.

estimated by 1.45 cwt. per acre. The ratio of the difference to its estimated S.D. is usually denoted by t:

$$t = \frac{3.0}{1.45} = 2.07.$$

If the true S.D. had been used, this ratio would be compared with the 0.05 significance level for the Normal distribution, but the effect of using only an estimate of the S.D. per plot is to make large deviations from zero rather more frequent. Sometimes the estimate of the S.D. will be larger than the true (unknown) value, and sometimes smaller, but, since the estimate is used as a divisor, the decreases in the t-ratio that occur when it is too large will be more than balanced* by the increases when it is too small. The consequences of this have been studied mathematically, and the theory is summarized by Table 5.5. As might be expected, if the number of plots is large the t-ratio will approximate to what would be found if the true S.D. could be used, but the smaller the number of plots is the more marked is its departure from Normality. Once more the key lies in the degrees of freedom. Table 5.5 shows the levels of t corresponding to probabilities of 0.1, 0.05, and 0.01 for various total numbers of d. f. The limiting values are identical with those already quoted for the Normal distribution in Chapter 4, but they increase rapidly for small numbers of d. f. At the extreme, if the experiment had two plots of A and only one of B, so giving 1 d. f., a ratio as large as 12.7 would occur by chance in 5 % of trials, instead of the ratio 1.96 for this probability in the Normal distribution. Table 5.5 is extracted from the much more extensive tables that have been calculated.

In the numerical example, the estimate of σ^2 used had 8 d.f. (4 from each variety), so that the 0.05 significance level for the t-ratio is 2.31. The calculated value, 2.07, does not exceed this. Hence the hypothesis of identical probability distributions for the two varieties is not contradicted by the evidence to any serious extent: the experimenter would be unwise to conclude that he had established an indubitable superiority for B.

---

* For in the present example suppose that the true S.D. for the difference in means were 1.60 cwt. per acre, so that the observed difference was 1.88 times its true S.D. Our experiment has underestimated the S.D. as 1.45 and has given t = 2.07; an experiment that overestimated the S.D. to the same extent (i. e. as 1.75) would give t = 1.72, which is not as far below 1.88 as 2.07 is above it. (This explanation is an oversimplification, but should be some help to the understanding of a difficult point).

The reader should make similar calculations for Tables 5.2 and 5.3; the t-ratios are 5.5 and 0.5 respectively, which, by reference to Table 5.5, are seen to agree with the conclusions drawn earlier after inspection of the data.

Table 5.5: Some significance levels for t.

Degrees of freedom	Probability 0.1	0.05	0.01
1	6.31	12.7	63.7
2	2.92	4.30	9.92
3	2.35	3.18	5.84
4	2.13	2.78	4.60
6	1.94	2.45	3.71
8	1.86	2.31	3.36
10	1.81	2.23	3.17
15	1.75	2.13	2.95
20	1.72	2.09	2.84
30	1.70	2.04	2.75
60	1.67	2.00	2.66
120	1.66	1.98	2.62
Limit	1.64	1.96	2.58

Significance and Estimation

To say that a difference is not statistically significant does not mean that it is necessarily small or insignificant in the colloquial sense. The experimenter who has obtained the data in Table 5.1, and who has made the analysis just described, should state that the difference in yields is non-significant. Nevertheless, if he is asked to estimate the difference between the true means for A and B, his best answer will be "3.0 cwt. per acre", an amount that is certainly not small and, if correct, might well be economically important.

The explanation of this seeming paradox is that statistical significance refers only to the weight of evidence, not to the magnitude of the effect. (A jury's verdict that evidence is insufficient to convict the prisoner is not always to be regarded as an assertion of their firm belief that he is innocent of even the smallest misdemeanour!) The experimenter may reasonably estimate one variety to be 3 cwt. per acre better than another, yet admit the lack of sufficient evidence to demonstrate satisfactorily that the difference is not zero.

†*Fiducial Limits*

Another way of presenting the conclusions from the analysis of the wheat experiment is often valuable, because it throws the emphasis on the magnitude of the difference in yield and avoids the awkwardness of statement discussed in the previous section. Analysis of Table 5.1 has led to 3.0 $\pm$ 1.45 cwt. per acre as the estimate of the difference in mean yield. As explained earlier, the mathematical theory of the distribution of t shows the probability that the estimated difference differs from the true difference by at least 2.31 times the standard error to be 0.05. Hence associated with the probability of 0.05 is a deviation 2.31×1.45, or 3.35 cwt. per acre. The statement may be made that the true difference in yields lies between 3.0—3.35 and 3.0+3.35 cwt. per acre, i. e. between 0.35 cwt. per acre in favour of A and 6.35 cwt. per acre in favour of B. If this statement is untrue, the experimental results are so extreme as to have a probability of less than 0.05, whereas if it is true the probability is greater than 0.05. This may be expressed by saying that there is a *fiducial probability of 0.95* that the true difference in yields lies between the two values quoted, and a fiducial probability of 0.05 that it lies outside that range; the values themselves are termed the *fiducial limits**. In careful logical discussion, the reader should avoid any idea that the true difference has in the ordinary sense a probability of 0.95 of being within the range of values determined. The true difference, though unknown to us, is unique in value and therefore cannot have a probability distribution; to distinguish our mode of thought from the more familiar relative frequency concept of probability, we introduce the adjective 'fiducial', and thus remind ourselves that the probability relates to the truth or falsity of our statements rather than to the true value of the quantity estimated. This subtlety may seem unimportant at a first reading, but can be a source of much confusion in more complex statistical argument and must therefore be mentioned.

In the test of significance, the experimenter might have chosen to make a different null hypothesis. He might have hypothesized that the yield distributions for A and B had the same standard deviations but differed in mean by 1.0 cwt. per acre in favour of B, or by 0.9 cwt. per acre in favour of A, or by any other amount he cared to specify. Some of these hypotheses would be contradicted by the t-test of significance at the 0.05 probability level, others would not. The fiducial limits are the most extreme values that just fail to be contradicted, as

* Some statisticians use *confidence limits* rather than fiducial limits. The logical relation between the two ideas is difficult to understand and is a source of much controversy. For most simple problems such as find a place in this book, the two types of limit are numerically identical and their practical interpretation is effectively the same whether fiducial or confidence theory is used.

may easily be verified by making the appropriate tests: each will give t equal to exactly 2.31. The experimenter may reasonably assert that the true difference is one that is not contradicted by a significance test, and that it must therefore lie between the fiducial limits*. Such a statement is not necessarily true, but, if he habitually makes statements by the same rules, about 95 % of them will be true: he is imperfect, but his standard of veracity is one that many might envy!

As for significance tests, fiducial limits for other probability levels may be calculated when required by using a t-ratio from a different column of Table 5.5. If the 0.01 probability is used instead of 0.05, the statements made will less frequently be untrue but this gain is paid for by a wider spacing of the limits so that the difference in yields is less precisely located; for Table 5.1, the statement would now be that the true difference almost certainly lies between 1.87 cwt. per acre in favour of A and 7.87 cwt. per acre in favour of B.

The concept of fiducial limits is confusing, especially when it must be explained without mathematics, and the reader should study the above paragraphs carefully. He should then verify that for the experiment reported in Table 5.2 the limits are 1.73 and 4.27 cwt. per acre in favour of B, and for Table 5.3 they are 11.0 cwt. per acre in favour of A and 17.0 cwt. per acre in favour of B. When a difference in yields is significant, both fiducial limits are on the same side of zero, but when it is non-significant the fiducial range includes zero.

†*Additional Information on σ^2*

Alert students often inquire why, in the analysis of data like those in Table 5.2, a sum of squares of deviations of all the ten values from their common mean is not used in the estimation of variance. The null hypothesis states that all come from the same distribution, and therefore apparently 9 d. f., rather than 8, should be available for this calculation.

The contention is sound, but its acceptance makes no difference to the test of significance. The ninth degree of freedom makes use of the difference between the A and B means (the reader may verify that it adds an amount $5\,(\bar{x}_B - \bar{x}_A)^2/2$ to the pooled sum of squares of deviations), and consequently the total sum of squares occurring in the numerator of s^2 is no longer independent of the difference between the means of the varieties. To allow for this requires a simple alge-

* Fiducial limits may be calculated in this way whether or not the difference in yield is significant, provided only that the assumption of equal standard deviations for individual plots of the two varieties is permissible. Usually this will be satisfactory; if not, other methods are required.

braic rearrangement of the significance test, which in fact leads to a probability level identical with that obtained by the method already described. Indeed, Table 5.5 could be modified so as to be appropriate to this kind of test, but for various reasons this is less convenient; the conclusions drawn from it would be identical with those from the present version.

The chief objection to using the data in this way arises when the verdict 'significant' requires the null hypothesis to be rejected. The assumption that the two varieties have the same S.D. is often tenable even when their means are clearly different, and a statement of fiducial limits may with advantage be based upon an estimate of this common S.D. The estimated S.D. must then be independent of the difference in means, and inclusion of the ninth degree of freedom would wrongly inflate it by a component due to a genuine varietal difference as well as to natural variation. Thus to form an estimate of σ^2 from all the data indiscriminately would not affect the test of significance, but would lead to a wrong assessment of fiducial limits.

Methods of Calculation

The method of calculating a sum of squares of deviations used in this example is the clearest for explanation of the process, but is rarely the best for practical computing. One of the commonest faults of the inexperienced user of statistical techniques is to form sums of squares in the way that Table 5.4 illustrates. Table 5.6 shows the alternative that is almost always quicker unless the numbers involved are unusually simple. Instead of calculating the individual deviations, the yields themselves are squared and added for each variety; from this total is subtracted the square of the total of the yields divided by the number of plots. The result is exactly equal to the sum of squares of deviations. This is a consequence of the algebraic formula

$$S[(x-\bar{x})^2] = S(x^2) - \frac{[S(x)]^2}{N}$$

for any number, N, of individual values.

The advantage for this method of calculation scarcely appears with these simple data; indeed the 'data' were artificially constructed to give an easy example in this chapter. The reader who is not convinced should look at Table 5.7, where the methods are compared for a hypothetical set of seven yields in which the individual deviations are not simple numbers. A third possibility is to subtract a constant amount (any amount!) from each value before squaring, in order to make the numbers to be squared much smaller. This also is illustrated in Table

5.7, and is often the best of all if no calculating machine is available since the number subtracted may be chosen as any convenient amount.

Table 5.6: Improved computation of sums of squares of deviations for Table 5.1.

	x_A	x_A^2	x_B	x_B^2
	19	361	23	529
	14	196	19	361
	15	225	19	361
	17	289	21	441
	20	400	18	324
Total	85	1,471	100	2,016
Adjustment	$85^2 \div 5 =$	1,445	$100^2 \div 5 =$	2,000
Difference		26		16

Table 5.7: Comparison of three methods of computing a sum of squares of deviations.

		Method I		Method II	Method III	
	x	$x-\bar{x}$	$(x-\bar{x})^2$	x^2	$(x-19)$	$(x-19)^2$
	23	2.29	5.2441	529	4	16
	19	−1.71	2.9241	361	0	0
	19	−1.71	2.9241	361	0	0
	21	0.29	0.0841	441	2	4
	18	−2.71	7.3441	324	−1	1
	24	3.29	10.8241	576	5	25
	21	0.29	0.0841	441	2	4
Total	145	0.03	29.4287	3,033	12	50
Adjustment				$145^2 \div 7 =$ 3,003.5714	$12^2 \div 7 =$	20.5714
Difference				29.4286		29.4286

CHAPTER 6

THE REDUCTION OF ERROR

"Errors like straws upon the surface flow;
He who would search for pearls must dive below."
Dryden: "All for Love", Prologue.

The Magnitude of the Standard Error

The experiment discussed in detail in the last chapter, and especially the consideration of the test of significance and the fiducial limits, should make the reader realize that a good experimenter will be concerned to keep his standard errors as small as possible — or rather, as small as practicable. The difference in true yielding capacity of two varieties of wheat is a property inherent in those varieties and in the soils, management and weather to which they are subjected; in any experiment, the difference observed will vary about the true value, and the standard error of that difference will depend upon the arrangement of the experiment. In an experiment with a large standard error, a true difference of, say, 3.5 cwt. per acre may fail to be detected because the observed difference is not significantly greater than the S.E. An experiment under similar conditions on the same two varieties, and therefore the same true difference, which has a small standard error may give a significant observed difference and so detect the true difference between the varieties. Naturally an experimenter always wishes to discover a true difference, if one exists, and so ought to take advantage of any legitimate methods for reducing his S.E.

Methods for the reduction of experimental error may be divided into six categories, according as they concern

(i) Character and size of experimental unit,
(ii) Replication,
(iii) Experimental technique,
(iv) Supplementary measurements,
(v) Experimental design,
(vi) Analysis.

Some of these are primarily the responsibility of the experimenter, others that of the statistician, but most need the cooperation of both. Each will be discussed briefly here.

The Experimental Unit

The reason for recommending that experiments on field crops have small plots, rather than whole fields, as the units of experimentation has already been stated (Chapter 5): plots within one field will vary less amongst themselves than will whole fields, and so will give smaller standard errors. Even if, on account of their small size, the S.D. per plot is as large or larger than would be the S.D. between whole fields, the possibility of using a much larger number of plots than fields will almost certainly ensure that the S.E. of a mean yield is smaller.

The question of the best size of plot has also been briefly mentioned in Chapter 5. Although the S.D. per plot will tend to increase as the size of the plot is decreased, because a smaller plot is necessarily less representative of the whole area of experimentation, the possibility of having a greater number of small plots usually makes them economically preferable. A lower limit to the dimensions is set by the need to provide, around the margin of each plot, guard areas which will be discarded from all harvest records. As explained in Chapter 5, this must be done in order to eliminate effects of competition between neighbouring plots. Thus the dimensions of the harvested area of a plot may be appreciably less than the total dimensions, and any attempt to use exceedingly small plots would involve great wastage of land on non-experimental guard areas. If experimentation on small plots is to be a trustworthy guide to the relative merits of various treatments or varieties when used on a field scale, under ordinary agricultural conditions, the areas whose yields are analysed must behave in the same way as would similarly treated whole fields. Fortunately the validity of inference from small plots is seldom impaired by the possibility that the treatments would show different effects when applied to whole fields; the interaction between neighbouring plots in respect of fertilizer or cultivation treatments, for example, can be satisfactorily removed by rejecting guard areas. Occasionally, however, effects on small plots are misleading: a small plot of a cereal variety that ripens early may suffer intensive bird damage because the varieties with which it is being compared ripen much later, whereas a whole field of the same variety would be proportionately less damaged.

For most agricultural experiments on annual crops, plots between

1/200 and 1/20 acre in area are usually desirable, but for grazing experiments or experiments on fruit or other tree crops substantially larger areas may be needed. Plots should be as alike in all important features (fertility, drainage, slope, etc.), as can be arranged*; they need not be internally homogeneous, provided that all have similar heterogeneity, as may occur for example when long narrow plots run down a slope from an area of good drainage to an area of bad drainage. Plots are almost always rectangular, but may be long and narrow rather than square if the agricultural operations to be performed on them makes this preferable. In all these matters, the experimenter must take the final decisions, but he will be wise to consult an experienced statistician who may have, or be able to find, information on the relative S.Ds. for different sizes and shapes of plot in previous trials of the same or a related crop. The whole experimental area should be reasonably homogeneous if this be practicable, but devices of design can do much to overcome difficulties under this head; markedly irregular and heterogeneous areas should be avoided.

Similar points arise in the choice of an experimental unit for other types of experiment. For work on animals, for example, the individual animal is usually the best unit. Mention has already been made in Chapter 2 of the danger of inadvertently making a whole flock of animals the unit; to keep animals in separate cages or enclosures may be difficult, but this trouble may be preferable to the large increase in S.D. that is likely if groups of animals are penned together. Thus, in experiments on pig nutrition, allocation of one or more whole litters to each diet under study may be convenient because the animals of a litter can be kept together. The experimenter must remember, however, that he may be able to obtain as small a S.E., and therefore as precise a comparison between two diets, by using fewer animals penned individually. The scheme adopted may have to be a compromise between what the statistician advises and what the experimenter insists, but the statistician's contribution should not be ignored. Occasionally the unit is a period in the life of an animal, as in a feeding trial that has alternative diets tested in successive periods on each of several animals; yet again, it may be a part of an animal, as in the comparison of types of tuberculin by measurement of the thickening of skin oc-

* In any one experiment, plots are generally identical in size and shape. When special features of the land available make it desirable, small differences may be permitted, but yields must then all be converted from lb. per plot to cwt. per acre (or other standard units) before the statistical analysis.

curring at the points of injection into the neck of a cow (the test being repeated on several cows). In any experiment, the animals should be as homogeneous in heredity and past history as the experimenter can secure.

A Note on Terminology

In general discussion of agricultural experimentation, the word *plot* is often used for any type of experimental unit, so giving a terminology for animal experiments as well as for field trials. The word *treatment* is used for any kind of factor whose different states are under study; it may relate to manurial, cultivation, or insecticidal treatments of crops, nutritional or other factors in the management of animals, or to different varieties of crop or breeds of animal. Most of this book is written in terms of experiments on field crops, but this is purely for convenience of phrasing and exactly the same statistical principles and techniques are applicable to animal experiments.

Replication

Experimenters often imagine that the first thought of a statistician is to demand far more plots than resources will permit. Admittedly, by repetition of the treatments to be compared on very large numbers of plots (that is to say, by *replication*), S.Es. could be made exceedingly small. Nevertheless, the assertion is misleading, because the statistician ought to be just as anxious as the experimenter to perform the experiment in an economical manner. Replication and plot size are closely related, since a reduction in plot size may allow the fitting of a greater number of plots into the area available. If the experimenter will state how large a difference between two treatments he thinks may be present, or how large a difference must be before he will regard it as important, the statistician ought to be able to judge the degree of replication (the number of plots of each variety or treatment) necessary in order to have a reasonable chance of detecting such a difference. The answer may be that the experiment would require far more plots than the experimenter can provide, but even the information that a satisfactory experiment is beyond his resources will be useful to him. On the other hand, the statistician may be able to show that, by proper choice of plot size and of design, adequate data can almost certainly be obtained with fewer plots than the experimenter imagined. That this second finding is the rarer is perhaps an indication of the optimism of experimenters rather than of the unreasonableness of statisticians. Another possibility (discussed more fully in Chapter 7) of immense practical importance is that of combining tests of different factors into a single experiment, so that one large study may replace several smaller ones at considerable net saving to the experimenter.

The same points arise in animal work, but the alternative experimental units are usually fewer and the individual animal is often a lower limit to the size of unit.

Experimental Technique

Defects of experimental technique ought never to be allowed to become a major source of experimental 'error'; indeed, the importance of having a carefully standardized and accurately executed technique ought not to need emphasis by the statistician. In any comparison of crop varieties, factors such as fertilizer applications, cultivation treatment, and seeding rates ought to be uniform over all plots, except when a factorial design (Chapter 7) requires their diversification in accordance with some predetermined scheme. Similarly, in a fertilizer trial no difference in the variety used ought to be allowed to affect conclusions. The experimental area should be accurately measured, and the plot boundaries correctly marked, so that no misinterpretation of yields may result from inequalities of plot areas. Not only should quantities of fertilizers be accurately weighed, but great precautions should be taken to ensure that the amounts intended for one plot are not by accident placed on another! Fertilizer applications, cultivations, or other agricultural operations should take place for every plot on the same day, or, if this be impossible, the work on any one day should complete a whole number of blocks of the design. At harvest, the produce of each plot should be distinctly labelled, as accidental interchanges or losses of certain plots may be impossible to correct. Every record should be made immediately on completion of the relevant observation; if subsequently a fair copy of the records is made, the original field notes should on no account be discarded, as they provide the only possibility of checking mistakes of copying. Especially is this desirable for records of yields or other numerical observations that form part of the results of the experiment. No investigator should trust himself to make faultless records of weights and measurements plot by plot in the field at the time of harvest, still less to perform correcly at that time any mental arithmetic necessary in order to produce totals for each plot. Every possible cross-check should therefore be imposed. In animal experiments, similar precautions are necessary, such as ensuring that diets whose nutritive values are to be compared should not differ in any factors other than those under investigation, guarding against the mixing and subsequent confusion of animals that ought to be differently treated, and protecting the animals from infections or any other influences that may introduce variation irrelevant to the subject of the experiment.

Much more could be written in detail on the technique of experimentation in agricultural research, but, although this concerns the statistician, it is not his responsibility. One final warning against unnecessary accuracy may be given: a sense of proportion should be retained in all measurements that are made. If the yield of a plot is of the order of 50 lb. and the S.D. per plot is expected to be about 5 lb., efforts to weigh the produce to the nearest ounce are quite wasted; uncertainty in the exact boundary of the area to be harvested will account for more than this. If the chemical constitutions of the produce of different plots are to be compared, the chemical determinations will almost certainly be made on small samples of produce. The errors in these analytical determinations may be much smaller than the S.D. per plot, but the benefit from this would be small. To take larger samples and to subject them to a less accurate but less laborious analysis may be a better policy, since it is likely to give fuller representation of the whole plots (Chapter 8). Although all these matters are primarily the responsibility of the experimenter, discussion with the statistician may be helpful especially if some numerical evidence on which to base the judgement is available.

†*Supplementary Measurements*

The S.E. appropriate to a comparison between two treatments can sometimes be reduced by taking account of supplementary measurements. For example, in an experiment on sugar beet, the stand might vary widely from plot to plot for reasons quite unrelated to the treatments on these plots (such as the flooding of a low-lying area of the field). If the number of plants on each plot is counted or estimated, an adjustment may be made to the yields so as to allow for this irregularity, and to compensate for a variation that has no relevance to the subject of the experiment. The most obvious way of doing this is to express the yields as weights per plant, but that tends to be too generous for the thinly populated plots, where surviving plants have had the opportunity of growing larger. A better procedure is to adjust by means of a regression calculated from an analysis of covariance. This method falls rather outside the scope of the present book, but is briefly described in Chapter 9. Another example is that of comparing the growths of animals subjected to different dietary treatments. The animals used will not be identical in weight at the beginning of the experiment, and the growth of any animal may depend upon its *initial* weight quite apart from the diet it receives. To regard the increase in weight over a fixed period as a measure of the effect of the diet can for this reason be misleading; the analysis of covariance is again a better procedure.

Experimental Design

By the design of an experiment is meant the logical relationships between the treatments to be compared and the set of rules for allocating treatments to plots.

The experiment that formed the subject of Chapter 5 was a very simple example of a *completely randomized* design. It had only two treatments, namely the two varieties; the rule of allocation to plots was that each variety should be grown on five plots, the division of the total of ten into two sets of five being made at random without any restriction. Restrictions on randomization may legitimately be introduced, provided that corresponding modifications are made in the statistical analysis. For example, in this experiment, the restriction might have been introduced that each variety was to be grown on one of each of five arbitrarily defined pairs of plots. Even a simple change of this kind may be important in increasing the precison of an experiment. Some of the more useful devices will be discussed in later sections of this chapter. In many experiments, more than two treatments are compared. These may be unrelated, as when six varieties are compared with one another simultaneously, or they may have certain inter-relationships, as when three varieties are compared both with and without nitrogenous fertilizer so giving a set of six treatments. Development of this possibility of factorial structure of the treatments is of great value to experimentation, because it increases the number of topics that can be investigated within the scope of the single experiment; Chapter 7 gives a brief introduction to this subject, to which statisticians have made many contributions in recent years.

Analysis

In general, all essential features of the statistical analysis of an experiment are determined by the design of the experiment. There are circumstances in which a statistician will deliberately employ an analysis that he regards as less efficient than the best possible, because he believes either that the data do not justify a fully efficient analysis, or that an efficient analysis would be so excessively laborious as to make the collection of more data preferable to its performance. Both these situations arise in statistical science, but both are rare in the standard type of agricultural experiment; the experimenter can seldom profitably ask the statistician to obtain a lower S.E. by an analysis more intensive and elaborate than that he has already made.

Most field experiments, if carefully designed, executed, and recorded, will have associated with them standard patterns of analysis which extract all the useful information. The rule for the experimenter should be either to rely upon his statistical colleague, or to make sure that he himself knows the exact form of analysis appropriate to his experiment and to apply it. He should particularly beware of spoiling good experimental technique by careless statistical analysis. A warning may seem unnecessary, but too often the inexperienced user of statistical techniques either tries to apply something that is not appropriate, or, still less excusably, spoils the correct method by careless mistakes of arithmetic. If he is careful of these points, the experimenter may rest assured that no new statistical analysis is going to show him how to obtain much smaller standard errors.

Rejection or Loss of Data

Experimenters sometimes wish to ignore the data obtained from one or more plots of an experiment which appear to be numerically either exceptionally large or exceptionally small; they may argue that the data cannot possibly be typical of the treatment to which they relate, or that they are so discrepant as to make too large a contribution to the estimate of S.D. Such rejection of data can rarely be approved, since it introduces a most undesirable element of personal judgement. If the data are known to be correct records of what was found in the field, then the exceptional yields have occurred in a properly randomized experiment, and should be regarded as genuinely representative of the wide variation that naturally occurs unless external evidence throws grave doubts on their authenticity.

Rejection of data can be justified, however, when the history of the suspect plots contains exceptional features that may be responsible and that cannot conceivably be consequences of the treatment of the plots. For example, if a plot at the edge of an experiment is damaged by stock, or is inadvertently in part harvested with a neighbouring commercial crop instead of with the experiment, however unfavourably these events reflect on the conduct of the experiment they do at least constitute good reason for rejecting the plot from the analysis. Again, a numerical record so extraordinary as to be beyond the bounds of credibility may demonstrate that a mistake of recording must have occurred: if the yield recorded for a plot of potatoes is found to be 46.4 tons per acre, or if a cow is recorded as having given 14.9 % butter-fat, no one would deny that the record should be disregarded.

Even these grounds for rejection need to be used with caution, however. To reject a yield from a plot of a cereal experiment because it has been reduced by bird damage may bias the judgement of a variety that ripens early and is therefore especially liable to attack by birds. The danger of rejecting data because the experimenter's preconceived ideas about the treatments make them seem unduly large or small rather than because they really surpass the bounds of credibility is too obvious to need further comment.

The remedy for most of these troubles lies in careful conduct of the experiment and accurate recording of measurements. Plots ought not to be damaged or lost from any preventable cause; all observations and measurements ought to be correctly made and carefully preserved. Even so, occasional losses (e. g. the death of an animal) or sound reasons for rejection will arise. In all but the simplest experimental designs, the loss of one or more plots complicates the statistical analysis, and the assistance of a professional statistician should be sought.

Randomized Blocks

If the reader has examined Fig. 5.1 carefully, he will have noticed a tendency for the plots of variety A to be at the top of the experimental area, and those of B at the bottom. Consequently, if there happened to be an increase in the fertility of the land from top to bottom, B will have been favoured relative to A, and the difference in yields will be greater than it ought to be. This does not mean that the experiment is unfair or biased. The complete randomization procedure ensures that, if the experiment were repeated many times with a fresh randomization for each, the varieties would have equal chances of the luck of position; the mechanism of the significance test, with its emphasis on the probability of 0.05 of calling a difference 'significant' even when it is entirely due to chance, ought to take account of the randomization arrangements that happen to be especially favourable to one variety. Nevertheless, the magnitude of the difference in yield observed would vary widely from one experiment to another because of variations in fortune at randomization.

The completely randomized design is therefore rarely used in field trials, although this design and the pattern of analysis described are often adopted for animal experiments and for many kinds of laboratory work. The disadvantage is usually overcome by introducing some form of *local control* into the design; the simplest form produces the *randomized block* design. Suppose that the ten plots are divided into five pairs, the two members of a pair being as alike in situation and fertility as the experimenter can judge; the pairs, or *blocks,* will usually be adjacent plots. From each block, one plot is selected strictly at random and assigned to A, the other being assigned to B. In this way, each variety is made to appear more uniformly over the whole area of the experiment. Fig. 6.1 shows how Fig. 5.1 might be modified to give a randomized block design. The five blocks consist of the pairs of plots I: 1,4; II: 2,5; III: 3,6; IV: 7,9; V: 8,10. One plot of block I is chosen at random for A; this might be done by spinning a well-balanced coin or by using the tables of random numbers men-

I	II	III
1 B	2 A	3 A
4 A	5 B	6 B
7 A	8 B	
9 B	10 A	
IV	V	

Fig. 6.1: Randomized block design for the comparison of two varieties.

tioned earlier. Thus perhaps plot 4 is assigned to A and therefore plot 1 to B. An entirely independent selection of a plot at random from block II is then made, and perhaps results in plot 2 having A, plot 5 having B. Similar selections complete the other blocks.

A change in the principle on which a design is constructed always requires that the form of analysis be changed also. The reader may at first find difficulty in understanding this. He will notice that the arrangement of varieties in Fig. 6.1 *could* have occurred as a result of complete randomization, and therefore might seem to need the same form of analysis. The arrangement in Fig. 5.1, however, could not occur as a randomized block design, since both plots of block I have variety A, both of block V have variety B. In fact, for this experiment, 252 different arrangements of the varieties could arise in complete randomization, whereas only 32 of these could also arise as randomized blocks. The form of analysis depends not on the particular arrangement of plots and varieties in the field but on the process of randomization that generated the arrangement, and hence on the set of arrangements from which the particular one was selected. The completely randomized design in Fig. 5.1 was chosen in such a way as to give all the set of 252 equal chances of being used, whereas the randomized block design in Fig. 6.1 gave equal chances to the smaller set of 32. The effect of this reduction in the freedom of randomization is to reduce the number of d.f. available for estimating the S.D. per plot;

in fact, of the 8 d.f. on which s^2 for the completely randomized design was based, 4 are removed as relating to variation between blocks and only 4 d.f. remain for the new s^2. If the blocks differ widely in fertility or in other qualities that influence yield, the new s^2 will be substantially smaller than that for a competely randomized design, and the test of significance of the difference between A and B will be correspondingly more sensitive.

When s^2 has been calculated, the formation of the S.E. for the difference in yields and the making of a test of significance follow exactly as in the completely randomized design. How is s^2 to be calculated? One method is a modification of that described in Chapter 5. A more generally useful method, however, is that known as *analysis of variance;* for an experiment with only two varieties, it leads to exactly the same conclusions, but it has the advantage of being easily adapted to experiments that test several varieties or treatments simultaneously. It will therefore be illustrated for a more ambitious experiment, still in randomized blocks but testing four varieties.

†*Systematic Designs*

Before describing the construction, analysis, and interpretation of a general randomized block design, we may consider briefly another type of design that would apparently be still more successful in eliminating variations of fertility over the experimental area. Some might raise an objection to the randomized block principle on the grounds that, for example, in Fig. 6.1 it permits a group of adjacent plots like 5, 6, 8 all to carry the same variety. Why should we not instead assign one of our two varieties, selected at random, to plot 1 and then let A and B alternate in a systematic pattern over all the plots? Fig. 6.2 shows such a design for the same experimental area; this is a type of design that was once popular with experimenters, who termed it a *checker-board* design.

If our sole aim in designing an experiment were to obtain the smallest possible S.E., this design would have many merits and might reasonably be chosen. To know that the difference in mean yields has practically the smallest S.E. attainable with a particular set of plots, however, is not necessarily the same thing as to be able to measure this S.E. and to use it in a test of significance. Although the checker-board design is one that could have occurred in complete randomization or in randomized blocks, it is now a random selection from only 2 possibilities (that in Fig. 6.2 and the same thing with A and B interchanged); therefore we must be prepared again to modify the analysis. Examination of the mathematical theory reveals that a further 4 d.f. are lost from the estimation of variance, so that now none remain! In other words, although the difference in mean yields is a valid estimate of the true relative values of A and B, and is likely to be more precise than the estimate obtained by using the designs in Fig. 5.1 or Fig. 6.1, we cannot calculate a S.E. for this difference. Consequently, the design is of severely limited utility, and should be employed only by an experimenter who is content to dispense with any assessment of S.E.

1 B	2 A	3 B
4 A	5 B	6 A
7 B	8 A	
9 A	10 B	

Fig. 6.2: Checker-board design for the comparison of two varieties. [See section on 'Systematic Designs' for explanation of why this design should not be used].

Many people find this point confusing and an analogy may help. If we propose to buy a clock from a firm of high repute, we may be content to accept their judgement of which is the best timekeeper at the price we are willing to pay, without demanding an exact statement of its accuracy; we may perhaps feel complete confidence that the clock will be accurate enough for any use we wish to make of it, and that the magnitude of the small errors to which it is subject is unimportant. On the other hand, if we have to buy a clock in a foreign country, where we do not know the reputations of clock-makers, we might reasonably decline to buy from a firm on the basis of their verbal statement of which was the best clock. We would perhaps prefer to buy our clock from another firm that was willing to provide a certification of accuracy, even though we might thereby obtain a poorer timekeeper than from the first firm: to know the greatest risk of error might be safer than merely to be assured that we had the best instrument available. The agricultural experimenter may sometimes be in the happy position of knowing his material so well that he can assess his standard error to a satisfying approximation without calculation from an experiment. More commonly, he is in a 'foreign country', and rightly demands that each of his experiments should provide its own estimate of S.E. He will therefore rarely be able to use checker-board designs (or similar systematic designs for a larger number of treatments) with any satisfaction.

Latin square designs, described in a later section, provide one method of ensuring a more thorough scattering of treatments over plots without sacrificing valid estimation of variance.

A Randomized Block Design

As an example of an experiment for the simultaneous comparison of several treatments, suppose that A, B, C, D, four varieties of wheat, are to be tested on an area that allows 20 plots to be used. Naturally, five plots of each variety will be grown, so as to give equal attention to each. A completely randomized design could be constructed by choosing five of the 20 plots entirely at random for A, then choosing five of the remaining 15 for B, then five for C, and the remaining five for D. This would be open to the objection raised earlier, namely that variations in plot fertility might contribute largely to the S.D., perhaps in a more acute form because of the larger area of the experiment. Instead a randomized block design would probably be preferred. One such is shown in Fig. 6.3. The five blocks there are the five rows of plots, but this is by no means always so: each block should be as homogeneous as possible, and the experimenter may use any knowledge of the site that he has in order to pick compact groups of plots likely to be similar in yielding potentiality. In this instance, one plot of block I was chosen at random (the third plot) and assigned to A, one of the remaining three was chosen at random (the second) and assigned to B, one of the remaining two assigned to C, and the last to D. A similar (but independent) randomization was then made for each of the other blocks. Fig. 6.3 also shows the yields, in cwt. per acre. In practice, the crop would be weighed as lb. per plot, the analysis would be performed in these units, and the final table of mean yields would be converted from lb. per plot to cwt. per acre by means of a factor determined from the dimensions of the plot; here, calculation in cwt. per acre throughout may help the reader to appreciate the meaning of the numerical values.

Although this book does not purport to teach methods of analysis, the analysis of variance is such an important technique that some account of it must be given. A first and obvious step is to rearrange the information on the field diagram, Fig. 6.3, into Table 6.1, in which the totals for each variety and the totals for each block are systematically set out. The analysis of variance is a method of apportioning the sum of squares of deviations of the individual yields from their general mean into components representing contributions from various sources. The meaning of these contributions depends upon the experimental design; their calculation for the randomized block experiment will be explained in the next section (which may be omitted by those who wish to avoid the labour of a little elementary arithmetic), but here the result will be shown. Table 6.2 shows the analysis of the total sum of squares of deviations into a component for differences between blocks, a component for differences between varieties, and a residual that for brevity is generally called *error*. The *mean square* obtained

Block I	C 25	B 23	A 20	D 20
" II	A 19	D 19	B 21	C 21
" III	B 19	A 14	D 17	C 18
" IV	B 19	D 13	C 21	A 15
" V	D 21	B 18	A 17	C 20

Fig. 6.3: Randomized block design for the comparison of four varieties, showing yields of wheat in cwt. per acre.

Table 6.1: Variety and block totals from Fig. 6.3.

Block	Variety A	B	C	D	Total	Mean
I	20	23	25	20	88	22.0
II	19	21	21	19	80	20.0
III	14	19	18	17	68	17.0
IV	15	19	21	13	68	17.0
V	17	18	20	21	76	18.0
Total	85	100	105	90	380	
Mean	17.0	20.0	21.0	18.0		19.0

Table 6.2: Analysis of variance for Fig. 6.3 and Table 6.1.

Adjustment for mean		7,220	Mean square
Nature of variation	d. f.	Sum of squares	
Between blocks	4	72	18.0
Between varieties	3	50	16.7
Error	12	36	3.0
Total	19	158	

from division of the error component by its degrees of freedom is s^2, the required estimate of variance per plot appropriate to the experiment. Proof lies beyond the scope of this book, but the discussion in the next section ought at least to make this seem plausible. The same technique of analysis could be applied to the completely randomized design discussed in Chapter 5, except that no component for 'blocks' would appear; it would give exactly the same value for s^2 as was there obtained. The interpretation of Table 6.2 is presented after the next section.

†*Calculation of the Analysis of Variance*

The general mean for the experiment is $380 \div 20$, or 19 cwt. per acre. Hence the sum of squares of deviations of all plot values from this, now taking no notice of variety, is (reading Table 6.1 row by row)

$$(20-19)^2 + (23-19)^2 + (25-19)^2 + \ldots + (18-19)^2 + (20-19)^2 + (21-19)^2 = 158.$$

The calculation is easily made here, because all the yields are simple figures and their mean is a whole number. Even so, the many small subtractions are tiresome and easily give rise to mistakes. Had the mean been a more awkward number, such as 19.78, the labour would have been much increased. A better procedure, leading to the same result, is that illustrated in Table 5.7: each of the 20 yield records is squared, and the sum of these squares is reduced by an *adjustment for the mean,* this being the square of the total of the 20 yields divided by the number of plots. The adjustment for the mean is wanted several times, and is therefore accorded a special niche in Table 6.2. The calculation, which the reader should complete for himself, is of the form:—

	400
	529
	625
	|
	|
	|
	|
	324
	400
	441
Total	7,378
Adjustment: $380^2 \div 20$	7,220
Difference	158

This sum of squares of deviations has been calculated from 20 yields and therefore has 19 degrees of freedom. It is entered in the last line of Table 6.2.

A component of the sum of squares attributable to differences between varieties may now be computed, by a similar operation on the total yields for the four varieties at the foot of Table 5.1. The calculation is

$$\frac{85^2 + 100^2 + 105^2 + 90^2}{5} - \frac{380^2}{20}$$

$$= \frac{36{,}350}{5} - 7{,}220$$

$$= 7{,}270 - 7{,}220$$

$$= 50.$$

The mathematical theory shows that the square of each varietal total must be divided by 5, because each is a total of 5 plots, and the adjustment for the mean is subtracted from their sum. There are 4 varieties, and the sum of squares corresponding to them therefore has 3 d.f. It is in fact the sum of squares of the deviations of varietal means from the general mean, 19.0, multiplied by 5 because each variety occurs 5 times:

$$5 \times \{(17.0-19.0)^2 + (20.0-19.0)^2 + (21.0-19.0)^2 + (18.0-19.0)^2\} = 50.$$

The first method of calculation, however, is the better for general use. The result is entered in Table 6.2 as the sum of squares 'between varieties'.

If the last sum of squares were subtracted from the total, the remainder would have 19—3, or 16 d.f., and would be equal to the total of the four sums of squares, each with 4 d.f., obtainable by the method used in the analysis of Table 5.1. Indeed, the error sum of squares for a completely randomized design, from which the estimate of the S.D. is calculated, is very conveniently obtained in this way. The reader should verify that, by such a calculation on the data of Table 5.1, he obtains a sum of squares of magnitude 42 with 8 d.f. He should also verify that the four sums of squares, each with 4 d.f., 'within varieties' in Table 6.1 are 26, 16, 26, and 40 for A, B, C, and D respectively; these have the total 108, which is 158—50.

In the randomized block design, however, comparisons between varieties are independent of block differences: since A and C both occur in every block, any tendency for blocks to differ in fertility will not affect the difference between the two varieties. The reader may like to verify that, if every yield in one block is altered by the same amount (say the four yields of block III are each decreased by 3 cwt. per acre), all varietal means are altered to the same extent and the differences between them are the same as in Table 6.1. The block classification is said to be *orthogonal* with the variety classification. A consequence is that a sum of squares 'between blocks' can be isolated, with 4 d.f., which is independent of that for varieties. The calculation is of the same kind as before, and either method may be used:

$$\frac{88^2 + 80^2 + 68^2 + 68^2 + 76^2}{4} - \frac{380^2}{20}$$
$$= 7{,}292 - 7{,}220$$
$$= 72.$$

The divisor, 4, is the number of plots per block. The result is entered in Table 6.2.

Subtraction of the two component degrees of freedom and sums of squares from the total line of Table 6.2 leaves a sum of squares of 36 with 12 d.f. This measures the residual variation after the block and variety contributions have been eliminated; since the estimate of variance per plot and subsequent estimates of S.Es. are derived from it, as already mentioned in the last section, it is termed the *error sum of squares*.

Interpretation of Analysis, and Conclusions

The greater the differences in mean yields for the four varieties, the greater will be the sum of squares between varieties, but the error sum of squares will not be affected by changes that apply to all plots of a variety: if D were replaced by a variety that yielded 5 cwt. per acre less on every plot, the sum of squares for varieties would be much larger, but that for error would remain the same. A suitable null hypothesis for the investigation of varietal differences is that the distribution of plot yields is the same for all varieties. On this hypothesis, not only is the error mean square an estimate of σ^2, the variance per plot, but also the mean square between varieties (here $50 \div 3$) will on an average be the same as this. Real differences between varieties will increase the mean square for varieties without affecting the other, and therefore a comparison of the two can give a test of significance. Once again, mathematicians have investigated the problem and have studied the probability distribution of the ratio of the two mean squares, popularly known as the *variance ratio* and denoted by F. They have prepared tables to show the level that would be exceeded by chance with various specified frequencies, on the assumption that the null hypothesis is true. The level for any probability depends upon the numbers of d.f. for both mean squares. Table 6.3 shows a selection of values for the 0.05 probability; similar but more extensive tables exist for this and other probabilities.

Table 6.3: Some 0.05 significance levels for the variance ratio, F

Error d. f.	Treatment d. f.					
	1	2	3	4	8	20
4	7.7	6.9	6.6	6.4	6.0	5.8
6	6.0	5.1	4.8	4.5	4.1	3.9
8	5.3	4.5	4.1	3.8	3.4	3.2
12	4.7	3.9	3.5	3.3	2.8	2.5
20	4.4	3.5	3.1	2.9	2.4	2.1
60	4.0	3.2	2.8	2.5	2.1	1.7

From Table 6.2, the variance ratio is

$$F = \frac{16.7}{30}$$
$$= 5.6$$

and the degrees of freedom are 3, 12. The significance level in Table 6.3, for these numbers of degrees of freedom, is 3.5; F is considerably greater, and provides ample justification for rejecting the null hypothesis. (The 0.01 level is 6.0, so that F does not quite attain this level). The experiment gives significant evidence of differences between varieties.

Merely to state that differences between varieties exist is inadequate, for the experimenter wants to know which varieties are good. Even though the null hypothesis has been abandoned, the S.D. per plot may be assumed the same for all varieties, and, as already pointed out, the error mean square,

$$s^2 = 3.0,$$

estimates the square of this. Table 6.4 summarizes the mean yields of the varieties, attaching to them the S.E. obtained by the usual square root rule, $s/\sqrt{5}$. Moreover, the S.E. of any difference between two varietal means is, by the rule of Chapter 5,

$$\sqrt{\left(\frac{3.0}{5} + \frac{3.0}{5}\right)} = \sqrt{1.20}$$
$$= 1.10.$$

To use this S.E. for examining differences between means of varieties would be unfair if the earlier test had shown no evidence of significant differences*. For the experiment in Fig. 6.2, however, the significance test has established that some real differences exist. If A is a standard variety, the experimenter will naturally wish to compare each of the others with it. Then, for B he will obtain a t-ratio

$$t = \frac{20.0 - 17.0}{1.10}$$
$$= 2.7,$$

well beyond the 0.05 probability level (Table 5.5); note that t has 12 degrees of freedom, the number for the error mean square. For D,

$$t = \frac{18.0 - 17.0}{1.10}$$
$$= 0.9,$$

which is nowhere near the significance level. The conclusion to be drawn is that varieties B and C are definitely better than the standard, but that D shows little evidence of any advantage; the difference between B and C is small and non-significant, but, if a variety must be chosen for future use, C seems the more promising.

Table 6.4: Mean yields for the experiment in Fig. 6.2.

Variety	A	B	C	D	S. E.
Yield (cwt./acre)	17.0	20.0	21.0	18.0	± 0.77

In Table 6.2, the mean square between blocks has been entered. A test of significance of this may also be made by forming its ratio to the error mean square and referring to Table 6.3 with 4, 12 d.f. The existence of significant differences between blocks is an indication that the experimenter has gained from using a randomized block design. Had he instead randomized completely over his 20 plots, this component would have been included with the error and we may estimate that the error mean square would then have been about 6.8.

---

* If each of a large number of pennies is spun 10 times, some may show tails in as many as 9 or 10 trials; this would not in itself be evidence that a few coins are biased because random causes alone will ensure that a proportion of perfectly balanced coins will behave in this manner. Similarly, a number of wheat varieties might have no real differences in respect of yield, yet in an experiment the one that chanced to be best might show considerably higher yields than the one that chanced to be worst.

7 ·

The experiment would therefore have been much less precise; the S.E. would have been larger, and differences in yield would have been less easily detected.

Latin Squares

The device of local control used for randomized block designs can be extended so as to eliminate variation in two directions simultaneously. In Fig. 6.3, the blocks take care of variation from top to bottom of the experimental area, which need not conform to any regular trend. Variations from left to right, however, will still influence the precision of comparisons if the randomization happens to give most plots of one variety on one side of the area; in this design, for example, most plots of B lie on the left and most plots of C on the right. In a *Latin square* design, the arrangement is balanced in both directions, an advantage that can be achieved if the number of replicates (the number of plots of each treatment) is equal to the number of treatments, and if the plots are in a square formation (though not necessarily square in outline).

Fig. 6.4 shows a Latin square design used in an experiment on the nitrogenous manuring of kale. All plots had dressings of potash salts and superphosphate. Plots denoted by the symbol A had no nitrogen, B and C had sulphate of ammonia to supply 0.14 and 0.40 cwt. nitrogen per acre respectively, while D and E had 0.40 cwt. nitrogen per acre from ammonium humate and humic acid respectively. The yields per plot are shown in the figure. For the first time, the data relate to a genuine experiment and are therefore less simple and less regular than in earlier examples! A Latin square design is such that each treatment occurs once in each row and once in each column, a property easily seen to be possessed by Fig. 6.3; for obvious reasons, the design of this experiment is called a 5×5 Latin square. Each row and each column contains all the five letters, and rows, columns, and treatments have the property of orthogonality mentioned earlier, in that anything affecting a whole row does not influence differences between columns or between treatments.

A Latin square for use in an experiment ought to be selected at random from all possible Latin square arrangements of the same size, just as is done for randomized block designs. In their book of statistical tables, to which reference has previously been made, Fisher and Yates have given simple instructions for doing this and special tables of squares to help the selection. Repeated use of one square in a series of experiments might result in neighbouring plots always having the same treatments, so biasing the comparisons that are made, and it is therefore to be strongly discouraged.

The statistical analysis of a Latin square is very similar to that for randomized blocks. It is described only very briefly here, for the benefit of those whose enthusiasm will persuade them to repeat the arithmetic for themselves. Table 6.5 contains totals of the kale yields according

D 740	A 560	E 580	B 600	C 620
E 750	B 570	C 660	D 600	A 450
A 730	D 600	B 600	C 700	E 450
C 760	E 600	D 580	A 530	B 490
B 820	C 720	A 550	E 520	D 560

Fig. 6.4: Latin square design for the comparison of five treatments.
A: No nitrogen
B: 0.14 cwt. N per acre as sulphate of ammonia
C: 0.40 » » » » » » » »
D: 0.40 » » » » » ammonium humate
E: 0.40 » » » » » humic acid
Yields of kale in lb. per plot.

to the rows of Fig. 6.4 (e. g. $740 + 560 + 580 + 600 + 620 = 3{,}100$), according to columns, and according to treatments. The analysis of variance in Table 6.6 is then formed just as was that in Table 6.2. Thus, for the total sum of squares,

$$740^2 + 560^2 + \dots + 520^2 + 560^2 - (15{,}340)^2/25 = 232{,}576.$$

For the component between treatments,

$$\frac{2{,}820^2 + 3{,}080^2 + 3{,}460^2 + 3{,}080^2 + 2{,}900^2}{5} - \frac{15{,}340^2}{25} = 48{,}736.$$

The numbers of d.f. are one less than the numbers of values for the corresponding categories, the sum of squares for error is the difference between the total and the other three components, and the mean squares are the results of dividing the sums of squares by the d.f. For the treatment and error lines of Table 6.6,

$$\begin{aligned} F &= 12{,}184/1{,}447 \\ &= 8.4, \end{aligned}$$

far in excess of the entry 3.3 for 4, 12 d.f. in Table 6.3. Differences between the fertilizer treatments are significant. Table 6.7 summarizes the mean yields in lb. per plot, as obtained from the last column of Table 6.5. The S.E. of these means is $s/\sqrt{5}$, where $s^2 = 1{,}447$. Each plot measured 30 ft. × 29 ft., an area of 1/50 acre, so that yields in lb. per plot must be multiplied by 50/2,240 to give yields in tons per acre; the same factor must be used for the S.E. The S.E. of a difference between two treatments is, in tons per acre, $\sqrt{2 \times (0.38)^2}$ or 0.54.

Table 6.5: Row, column, and treatment totals from Fig. 6.4.

Rows	Columns	Treatments
3,100	3,800	A: 2,820
3,030	3,050	B: 3,080
3,080	2,970	C: 3,460
2,960	2,950	D: 3,080
3,170	2,570	E: 2,900
15,340	15,340	15,340

Table 6.6: Analysis of variance for Fig. 6.4 and Table 6.5.

Adjustment for mean		9,412,624	Mean square
Nature of variation	d. f.	Sum of squares	
Between rows	4	4,936	1,234
Between columns	4	161,536	40,384
Between treatments	4	48,736	12,184
Error	12	17,368	1,447
Total	24	232,576	

Inspection of Table 6.7 then makes clear that the kale responded well to sulphate of ammonia, the increase in yield for the larger dressing being 2.8 ± 0.54 tons per acre. This effect is mainly responsible for the significant differences between treatments, but both the smaller dressing of sulphate of ammonia and the ammonium humate gave yields significantly higher than that without nitrogen. Humic acid had no significant effect on yield.

Table 6.7: Mean yields for the experiment in Fig. 6.4.

Units of yield	No nitrogen	Sulphate of ammonia 0.14 cwt. N	Sulphate of ammonia 0.40 cwt. N	Ammonium humate 0.40 cwt. N	Humic acid 0.40 cwt. N	S. E.
lb. per plot	564	616	692	616	580	17.0
tons per acre	12.6	13.7	15.4	13.7	12.9	0.38

Animal Experiments

The designs that have been described for field plot trials are also useful for animal experiments and for many laboratory purposes also. Location on the ground is replaced by other natural groupings of experimental units. For example, randomized block designs are valuable when several litters of animals must be used in an experiment; provided that the number of treatments to be compared is less than the size of the smallest litter, each treatment is assigned to one animal chosen at random from each litter (surplus animals being discarded from the experiment). In this way, the measurement of the relative effects of treatments is freed from a large part of the variation between animals attributable to their differences in heredity. Randomized block experiments on pigs have often been arranged on this basis. The design is especially useful when genetically identical animals can be used as blocks. In Sweden, identical-twin cattle have been collected from all parts of the country, by offering their owners special prices, and have been used in nutritional and other studies. A similar scheme in New Zealand is reported to have led to a fifteen-fold increase in precision for experiments comparing butter-fat percentages, to a twenty-fold increase for comparisons of milk yield, and to notable gains (sometimes greater and sometimes less) for comparisons of other measurements that interest experimenters; this means, for example, that a comparison between the milk yields obtainable on two alternative diets could be as precise if made on five pairs of identical twins as if it were made on two groups of 100 unrelated animals.

Latin squares have also found uses in animal experimentation. One important example is the *cross-over* or *change-over* type of experiment in nutritional studies. If five rations were to be compared, each column of Fig. 6.4 might be made to correspond to a different animal and the succession of letters in the columns would indicate periods on each ration; thus, the third animal would begin on ration E, would later be changed to C, later again to B, and so on. Measurements would be made on each animal during or at the end of each period. Latin squares of size 3×3 and 4×4 are especially suitable for this kind of experiment. By using several different squares in one experiment, residual effects of rations in previous periods can be balanced out or separately estimated. Table 6.8 shows a scheme that was used in an experiment on the feeding of dairy cattle; each cow received each of three rations for six weeks, and the merits of the rations were compared in terms of total milk production per period. The design not only ensured that six of the eighteen cows were on each ration in any one period but also secured a balance of residual effects: in Period 2, three of the cows on ration a had had ration b in Period 1 and three had had ration c, and similar restrictions held for other rations and other periods.

Table 6.8: Sequence of treatments in a cross-over experiment on the feeding of dairy cattle.

Period	Group I			Group II		
	No. 1	No. 2	No. 3	No. 4	No. 5	No. 6
1..........	a	b	c	a	b	c
2..........	b	c	a	c	a	b
3..........	c	a	b	b	c	a

Four other groups of three cows were used, Groups III and V following the scheme of Group I, Groups IV and VI following that of Group II

a: Hay and silage alone
b: Addition of 1 lb. grain per 7 lb. milk produced
c: Addition of 1 lb. grain per 3½ lb. milk produced.

†*"Yields"*

The statistical procedures that have been illustrated for wheat yields can be applied to any other continuous variate measurable on experimental units, such as heights of plants, dates at which a specified stage of growth is reached, milk yields, butter-fat percentages, growth rates, or lengths of life. They can often be applied to measurements of a discrete variate, especially to fairly large frequencies. For example, if insecticidal treatments are being compared in a field trial, counts of surviving insects might be subjected to an analysis of variance provided that they were all fairly large*; on this question, however, the advice of a statistician should be sought in each new instance.

†*Incomplete Blocks*

If the number of treatments to be tested in an animal experiment exceeds the litter size, or the size of any other appropriate set of blocks, more complex *incomplete block designs* must be used. Statisticians have exercised their ingenuity in order to retain as much sym-

* As emphasized in Chapter 2, the χ^2 test for a contingency table must never be used on measurements of a continuous variate; to use methods devised for a continuous variate on frequency data, however, is likely to be at worst a sacrifice of information unless the frequencies are very small.

metry as is possible, but the calculations become more difficult than those described here. In field experiments also, a limit is set to block sizes by the consideration that the plots within a block ought to be fairly homogeneous. In practice, field experiments with blocks of more than 12 plots or with Latin squares larger than 9×9 are rarely advisable, and a statistician can usually suggest an improvement. The *balanced incomplete block, partially balanced incomplete block,* and *lattice* designs are sufficiently numerous and varied in character to permit the inclusion within a single experiment of large numbers of varieties or treatments for comparison (even 100 or more can be managed), without employing excessively large blocks. All these are schemes for dividing the whole set of treatments into smaller groups that may be used as blocks; the division follows some symmetrical or semi-symmetrical plan, and usually pairs of treatments occur together in some blocks but separately in others. Whatever the rule for allocating treatments to blocks, the order in which they are placed on plots within a block must be strictly random. Enough has been said here, however, to introduce the reader to ideas of experimental design, and he must seek fuller information in more specialized books.

Presentation of Results

An analysis of variance is part of the statistical machinery for interpreting an experiment and is not itself a summary of the experiment. Writers of reports on agricultural researches sometimes present an analysis of variance and a comment that treatment differences are (or are not) significant as their complete statement of conclusions from an experiment. This is quite inadequate.

A proper statement of conclusions will always contain a table of mean yields (like Tables 6.4 and 6.7), or several tables if the experiment is more complicated; this should be followed by a brief discussion of the inferences that may be drawn, in terms of agricultural meaning and not merely in statistical jargon. Whether the analysis of variance itself should also be presented is largely a matter of taste. In any publication intended to explain methods of statistical analysis or to draw attention to difficulties of statistical technique, it will naturally find a place. In elementary reports intended to put purely agricultural information before the reader, it is quite inappropriate. In the standard type of paper for a scientific journal, it is rarely essential but may reasonably be given if the author finds that its presence helps his discussion.

CHAPTER 7

FACTORIAL DESIGN

> "... find out the cause of this effect,"
>
> *Shakespeare:* "Hamlet", Act II, Scene 2.

> "The more facts one has, the better the judgment one can make, but one must never forget the corollary that the more facts one has, the easier it is to put them together wrong."
>
> *Sir Geoffrey Heyworth:* Inaugural Address as President of the Royal Statistical Society, 1949.

The statistician's contribution to reducing the standard errors of experiments by the elaboration of experimental designs has been described in Chapter 6. He has made possibly an even greater contribution to the science of experimentation by the development of *factorial designs.* Not only do these permit tests of several different questions to be made simultaneously, with little additional labour, so effectively achieving the same gain as a reduction in the standard deviation; they also increase the scope and utility of the results, by providing information on the interactions between the factors under test.

Three Quotations

"The points at issue may be made clear by the consideration of an example. Suppose it is desired to introduce a new crop into a country, and that nothing is known of the most suitable varieties, type and quantity of manuring, the best cultivations, etc. The classical procedure would be to set up separate experiments to determine the best varieties, others to investigate the manurial requirements, others (if indeed any were undertaken on this point) to determine the most suitable methods of cultivation, rates of sowing, etc. Unfortunately, however, we cannot conduct manurial experiments without choosing some variety on which to conduct them, nor can we conduct varietal trials without deciding on some level of manuring, a rate of sowing, width between rows, cultivations, etc. Now it may happen that the effects of fertilizers on the different varieties are materially different, or that varieties that are good yielders at wide spacings, owing to a rank habit of growth, are much inferior in yield (or other qualities) when sown at close spacings. Thus conclusions that have been laboriously reached on the correct level of manuring for one variety may be inapplicable to the variety finally chosen, and that variety may itself be incorrectly chosen through not realizing the possibilities of increasing the yield of other varieties by changes in cultural practices."*

"If all combinations of two or more contrasts are tested factorially all plots contribute some information on each of the factors tested. If the response to one treatment is but slightly affected by variations in other treatments then the finding may be expected to prove applicable over a wider range of conditions than if it had initially been tested under only one rather arbitrarily chosen set of conditions. If, however, the response to one treatment is profoundly modified by some other treatment, then much will be learnt about the conditions under which response to the first treatment may be expected to recur in practice. Thus, if, as is general, liming reduces the response to mineral phosphate, one has fairly strong presumptive evidence that it would be unwise to recommend mineral phosphate for neutral soils. If, as is usual in Britain, on plots with dung the responses to phosphate are reduced to about one-half and those to potash are reduced to one-third of those on similar plots without dung, this is evidence that the benefit from dung must be in large measure due to the amounts of available phosphate and potash it supplies. Indeed, factorial experiments testing combinations of dung or compost with individual fertilizers provide one of the few ways in which the manifold possible effects of dung can be analysed into simpler terms. Quite often the effects of varietal or cultural treatments are profoundly modified by differential manuring. If two varieties give similar results over a wide range of manurial conditions, they are more likely to give similar results over a wide range of soils. If two cultural treatments give different results on manured and un-manured land, one has learnt something vital about the dominant factors in soil fertility and husbandry under local conditions."**

"No aphorism is more frequently repeated in connection with field trials, than that we must ask Nature few questions, or, ideally, one question at a time. The writer is convinced that this view is wholly mistaken. Nature, he suggests, will best respond to a logical and carefully thought out questionnaire; indeed, if we

---

* From *The Design and Analysis of Factorial Experiments,* by F. Yates (1937: Imperial Bureau of Soil Science); this and *Experimental Designs* by W. G. Cochran and G.M. Cox (1950: John Wiley & Sons) are two of the most valuable treatises on the technique of experimental design, but both are considerably more advanced than my book.

** From *Fertilizer Experiments in Colonial Agriculture,* by E. M. Crowther (1947: His Majesty's Stationery Office), a short and elementary but useful survey of the more important types of design employed in agricultural research.

ask her a single question, she will often refuse to answer until some other topic has been discussed."*

The Need for Factorial Designs

As the above quotations indicate, factorial designs are valuable in every type of agricultural research. For example, in a study of the effect of restricting the light received by a certain species of plant, the experimenter may plan to compare plants grown under normal lighting with plants whose light is cut down to two-thirds of normal and with others having only one-third of normal lighting. The effect of light is likely to be dependent in some degree upon the nutrition of the plant, and an experiment performed entirely at one level of fertilizer application (or entirely without fertilizer) may be seriously limited in its interpretation. A great improvement will be made if the three treatments are increased to six by testing each of the three levels of light intensity both with and without nitrogenous fertilizer. If the differences between the three levels of light intensity are practically the same with and without nitrogen, then the basis of the inference to be made from the experiment is broadened and strengthened. If the differences are larger (or smaller) when nitrogen is given than they are without, a valuable new fact has been discovered by the experiment, a fact that no amount of experimentation at one fixed level of nitrogenous fertilizer could ever have disclosed.

This experiment is said to have a *3×2 factorial scheme,* since one factor (light) is tested at 3 levels (1, 2/3, 1/3) and the other (nitrogen) is tested at 2 levels (absence and presence). The full set of six treatments might be compared by any of the experimental designs mentioned in Chapter 6. For example, five randomized blocks of 6 plots each, or a 6 × 6 Latin square might be used. The randomized block design would assign 10 of its 30 plots to each of the three levels of light intensity; 5 of each set of 10 would be without nitrogen and 5 with nitrogen, so that differences between average measurements at the three light intensities would be balanced in respect of presence and absence of nitrogen. In fact, for the purpose of comparing light intensities, the experiment would be almost as good as one using the same number of plots for that purpose alone. Moreover, the factorial experiment would assign 15 plots to each of the nitrogen treatments; of the 15 plots with nitrogen, 5 would have each of the levels of light intensity, and of the 15 plots without nitrogen the same would be true, so that any differential responses to light would apply equally to both

* From "The arrangement of field experiments", by R. A. Fisher (1926, *Journal of the Ministry of Agriculture*, vol. 33, 503—513).

nitrogen treatments. Hence, for assessing the effect of nitrogen, the difference in means would be balanced in respect of light intensity, and the experiment would be almost as good as one of the same size for this purpose alone. In addition, the factorial design provides information on whether, and to what extent, the effects of light intensity depend upon the presence or absence of nitrogen; this is known as the *interaction* between the effects of light and nitrogen, and may alternatively be regarded as showing how the magnitude of the response to nitrogen depends upon the light intensity.

The qualification 'almost' has been inserted twice in the last paragraph, because an experiment on light intensity alone could use blocks of 3 plots instead of blocks of 6, and an experiment on nitrogen alone could use blocks of 2 plots instead of blocks of 6, but the gain from inclusion of the additional factor is likely to outweigh by far the disadvantage of the larger blocks. Fuller information on the effect of nitrogen and its interactions could be obtained by using a 3×3 factorial scheme with zero, single, and double amounts of nitrogenous fertilizer. Moreover, the experiment could be made still more comprehensive by including combinations with various amounts of other fertilizers.

One of the most important uses of factorial designs is in the study of fertilizer requirements. If the responses of a particular crop to nitrogenous, phosphatic, and potassic fertilizers were to be investigated by means of separate experiments, a minimum requirement would be three experiments, each of perhaps 8 randomized blocks of 2 plots (each block having a plot without and another with the fertilizer under test), a total of 48 plots. Each experiment would have to be conducted at an arbitrarily chosen level of the other two fertilizers. None of the experiments would give information on whether responses to nitrogen were independent of, or were modified by, the presence of phosphate and potash. If instead a set of eight treatments is formed, these being all combinations of absence or presence of each fertilizer, the same total number of plots can be used with a design in 6 randomized blocks of 8 plots. Average differences in yield for plots with and without nitrogen will now be measured from two sets of 24 plots; although the larger blocks are likely to make the S.D. greater than for the experiment on nitrogen alone, the improvement in precision resulting from the greatly increased replication will almost certainly more than balance that loss. The same precision is obtained for estimating the responses to phosphate and to potash. In addition, three two-factor interactions can be examined, giving answers to questions such as "Is the response to nitrogen (i. e. the increase in yield associated with application of the nitrogenous fertilizer) increased, decreased, or unaltered by the presence of potash?" or, an equivalent question, "Is the response to potash modified by the presence of nitrogen?" Finally, a three-factor interaction can be examined: "Is the magnitude of any interaction between the effects of nitrogen and potash modified by the presence of phosphate?"; this last, and two symmetrical forms of the same question, are usually less important and are more difficult to interpret, but may sometimes be useful. The arrangement of all combinations of three factors, each at two levels (absence and presence) is described as a $2 \times 2 \times 2$ or 2^3 factorial scheme.

The second of the quotations in the previous section has drawn attention to the importance of factorial experiments in evaluating the manurial properies of dung. Only by study of the extent to which the increase in yield associated with an application of dung is modified by the presence of artificial sources of nitrogen, phosphate, or potash, can the availability of these nutrients in the dung be assessed: a factorial design with two or more levels of dung and two or more levels of each artificial is needed.

The examples above have involved quantitative factors — relative amounts of light or quantities of fertilizer. Factorial schemes are equally applicable with qualitative factors, such as different forms of cultivation, different types of phosphatic fertilizer supplying equal quantities of P_2O_5, different varieties of seed, and so on. The different states of any one factor are often termed *levels,* whether the factor is quantitative or qualitative.

Notation

The treatment combinations in a factorial design are usually represented by groups of letters (small rather than capital). For example, in a 2^3 experiment in which the factors were absence or presence of nitrogenous, phosphatic, and potassic fertilizers, the symbol nk would be used to represent the treatment consisting of application of nitrogen and potash but, by implication, no phosphate. The full set of eight treatments would be:—

(1), n, p, np, k, nk, pk, npk;

the symbol (1) is used for the zero level of all factors (here absence of all three fertilizers), rather than 0 or (—), for reasons that will not be explained here. So in a $3^2 \times 2$ experiment in which the factors were three levels of dung, three levels of nitrogen, and two dates of sowing, the two upper levels of dung could be represented by d_1, d_2, the two upper levels of nitrogen by n_1, n_2, and early sowing by e. Then (1) would indicate the lowest level (possibly zero or complete absence) of dung and nitrogen with late sowing, e would indicate the same manuring but early sowing, d_1n_2e would indicate the middle level of dung, the upper level of nitrogen, and early sowing, and the full set of treatments could be written in systematic order:—

$$(1),\ d_1,\ d_2,\ n_1,\ d_1n_1,\ d_2n_1,\ n_2,\ d_1n_2,\ d_2n_2,$$
$$e\ ,\ d_1e,\ d_2e,\ n_1e,\ d_1n_1e,\ d_2n_1e,\ n_2e,\ d_1n_2e,\ d_2n_2e.$$

The lowest level of a factor, expressed by absence of the relevant letter, need not be actually a zero quantity, although in practice it often is. Thus, in this last experiment, if the three levels of dung were to be 5, 10, and 15 tons per acre, absence of the letter 'd' would denote 5 tons per acre. Nor need the levels correspond to any ascending scale of values. If three entirely different forms of cultivation are to be compared, two will be represented by c_1, c_2 and the third by the absence of the letter 'c'; the order of allocation of the cultivations to these symbols would not matter, though if one were to be regarded as a standard it would most conveniently be represented by absence of the letter.

Variants of this system of notation are preferred by some people, especially by those who wish their symbols to be more directly suggestive of their meaning. For example, in the $3^2 \times 2$ design described above, they might write d_0n_0e rather than e and d_0n_0l rather than (1); for three distinct forms of cultivation, they might use different letters, a, b, and c, rather than c_0, c_1 and c_2. The system adopted here has advantages in that it facilitates the process of constructing such complex designs as those mentioned in later sections, since application of certain simple rules (not stated here) to the symbols discloses important characteristics of a design. Hence the statistician will usually choose this notation; if the experimenter wishes to identify his combinations of treatments in some other way that is clearer to him, he may easily do so provided that care is exercised in translation from one system to the other.

The capital letters corresponding to the treatment symbols, are used to represent numerical measures of the effects of treatments. In the 2^3 experiment mentioned above, N would represent the *main effect* of nitrogen, the response to nitrogen averaged over all levels of the other factors; similarly P would represent the main effect of phosphate. In the $3^2 \times 2$ experiment, N would again represent the main effect of nitrogen (now to be understood as the set of differences between the three levels of nitrogenous manuring and no longer a single number), end E would represent the mean advantage for early sowing relative to late sowing. Groups of capital letters represent interactions. Thus NP (sometimes written $N \times P$ or N . P) is a numerical measure of the interaction between nitrogen and phosphate for the 2^3 experiment and DNE (or $D \times N \times E$ or D . N . E) is a set of numerical measures of the three-factor interaction in the $3^2 \times 2$ experiment. The exact sense of these measures of interaction can scarcely be understood by those who have not mastered the methods of statistical analysis appropriate to factorial designs, but all should be able to appreciate the general idea.

Types of Design

In theory, a factorial experiment could contain any number of factors at any number of levels. For example, an experiment on potatoes might be planned to contain all combinations of five varieties of seed, from virus-free and from non-virus-free stocks, planted on three different dates, on land either shallow or deep ploughed, with three different rates of application of dung, four rates of sulphate of ammonia and two rates each of superphosphate and sulphate of potash, in all a $5\times2\times3\times2\times3\times4\times2\times2$ or, more concisely, a $5\times4\times3^2\times2^4$ factorial scheme involving 2,880 treatment combinations! In practice, considerations of space and of other resources limit the number of treatments to a much smaller figure and, at any rate in agriculture, experiments with more than 256 ($= 2^8$) treatment combinations are very rare. Moreover, the device of 'confounding', which is described below, makes desirable (though not essential) an arrangement in which every factor has the same number of levels. In particular, designs with several factors each at two levels (e. g. 2^3, 2^4, 2^5) or each at three levels (e. g. 3^3, 3^4) are very widely used, and are generally preferable to designs for mixed factorial schemes such as 3×2^2 and $3^2\times2^2$. No complete proscription of the mixed schemes is intended, for sometimes the aim of the experiment makes one of them necessary. Often, however, an experimenter who is considering a $3^2\times2$ scheme will be happy to introduce additional levels of the third factor and to have instead a 3^3 scheme; especially is this so when the statistician tells him that a design for the first in 36 plots may be little more precise than one for the second in 27 plots, yet almost certainly will be more difficult to analyse and interpret. Factors with more than three levels are more rarely used, though they are occasionally wanted; a factor at four levels can sometimes be regarded as equivalent to two factors at two levels each ($4 = 2\times2$), so simplifying the structure of the design.

When the factorial scheme has been chosen, the experimenter knows the total number of treatments that must be tested (e. g. for a 3^3 scheme, $3 \times 3 \times 3$ or 27 treatments); any design for this number of treatments may then be considered for use. Fig. 7.1 shows an example of a 2^2 scheme arranged in a 4×4 Latin square. Unless the number of factors is very small, however, the most obvious designs involve blocks larger than is generally satisfactory for work on field plots;

s	ks	(1)	k
k	(1)	ks	s
ks	k	s	(1)
(1)	s	k	ks

Fig. 7.1: Plan of a 2^2 experiment on sugar beet, in a Latin square.
Muriate of potash: None, or 0.8 cwt. Cl per acre (k)
Agricultural salt: None, or 0.8 cwt. Cl per acre (s)
Plot size: 1/60 acre.

only schemes like 2^2, 2^3, 3^2, 3×2, 3×2^2, 3^2, and perhaps 2^4 can be arranged in randomized blocks or Latin squares without serious danger of large standard deviations arising from heterogeneity within blocks or within rows and columns. Incomplete block designs must often be used. The symmetrical types, such as are valuable in tests of large numbers of varieties, may be chosen, but for an experiment of factorial structure the device of confounding is more commonly adopted because it enables greater precision to be obtained on the most important comparisons.

†*Confounding*

In a field experiment, considerations of labour or of space generally limit the number of plots that can be used; the number and shape of any blocks, however, can usually be adjusted to suit the needs of the experiment, since they are determined merely as lines drawn on the field plan. Now the S.D. of plot yields will tend to be increased by any increase in block size, and therefore the experimenter will wish to keep his blocks small. Apparently a lower limit to size will be set by the number of treatment combinations that the nature of his investigation obliges him to include, yet this can be achieved only by sacrificing other factors of lesser interest that are none the less relevant. Even the minimum number of combinations may exceed what is commonly regarded as the upper limit of reasonable block sizes. The conflict between precision (as represented by a small S.D.) and comprehensiveness (as represented by inclusion of many factors) can be resolved only by using blocks in which the number of plots is less than the number of treatment combinations. Animal experiments give rise to the same problem, sometimes in even more acute form since the possible block size may be more rigidly determined and perhaps much smaller than in a field experiment. In Chapter 5, the advantages to be gained from using monovular twins in cattle to give randomized blocks of two units (or 'plots') were mentioned, yet this may seem impossible unless only two treatments are to be compared.

When a factorial system of treatments is to be tested, judicious choice of the constitution of *incomplete blocks* (i. e. blocks containing less than the total number of combinations) may enable all the advantages of smaller blocks to be obtained without serious impairment of the interpretation of the experiment. A simple example will illustrate how this can be done, although the full merits of the method scarcely appear until more complicated designs are considered. If a 2^3 factorial scheme for factors n, p, k is to be arranged in ordinary randomized blocks, each block must have 8 plots. Suppose instead that blocks of 4 plots are used, that half the blocks have the treatments

(1), n, k, nk

randomized over their plots, and that the other half have the treatments

p, np, pk, npk.

Then clearly the effect of phosphate (symbolized by P) can be estimated only by the difference between the two types of blocks, since one type has all its plots with phosphate and the other is entirely without phosphate. The effect of phosphate is said to be *confounded* between blocks: any increase in yield caused by the phosphatic fertilizer cannot be distinguished from overall differences between blocks having no connexion with treatments. On the other hand, the effect of nitrogen (N) can be estimated from differences within blocks just as if the full blocks of 8 plots had been used. Each block has two plots without nitrogen and two having the same phosphate and potash treatments as these but with nitrogen: the nitrogen effect is not confounded. Similarly the potash effect (K) is free from confounding. Moreover, the interaction between the effects of nitrogen and potash (symbolized by NK) is unconfounded, being independent of block differences: it is estimated from the difference between the effect of nitrogen in the absence of potash and the effect in the presence of potash which difference involves two plots positively and two negatively from each block*. Similarly, the NP, PK, and three-factor or NPK interactions are unconfounded.

The design confounded as above is not much practical use, because it sacrifices information on the main effect of phosphate, one of the

Block I	np	(1)	pk	nk
» II	n	npk	k	p
» III	k	p	npk	n
» IV	(1)	pk	nk	np
» V	n	p	k	npk
» VI	pk	(1)	nk	np

Fig. 7.2: Plan of a 2^3 experiment on potatoes, confounded in randomized blocks of four plots.
Sulphate of ammonia: None, or 0.6 cwt. N per acre (n)
Superphosphate: None, or 0.6 cwt. P_2O_5 per acre (p)
Sulphate of potash: None, or 1.5 cwt. K_2O per acre (k)
Plot size: 1/60 acre
Confounding: NPK.

more important features of the experiment. If instead the two types of block had the treatments

(1), np, nk, pk

and

n, p, k, npk,

the reader may verify that the three main effects, N, P and K, and the three two-factor interactions, NP, NK, and PK (measuring the

---

* In the first type of block, the difference between plots receiving treatments n and (1) measures the effect of nitrogen in the absence of potash and the difference between treatments nk and k measures the effect of nitrogen in the presence of potash. Hence

$$\{nk - k\} - \{n - (1)\}, \text{ or } nk - n - k + (1)$$

is a measure of the interaction. Similarly, in the second type of block, the two forms of nitrogen effect are measured by the differences between treatments np and p and between treatments npk and pk, so that

$$\{npk - pk\} - \{np - p\}, \text{ or } npk - np - pk + p$$

is a measure of the interaction. The average of these two quantities gives the final measure of interaction that is adopted.

extent to which the response to each fertilizer is modified by the presence of one of the others) are all unconfounded. Only NPK is confounded, and this is a much less important loss, because experience shows high-order interactions often to be small. Indeed, even when high-order interactions exist, their practical interpretation and use is

Block				
Block I	p	nk	dnp	dk
» II	dnk	np	dp	k
» III	n	d	dnpk	pk
» IV	dpk	npk	(1)	dn
» V	dpk	dnp	nk	(1)
» VI	dnpk	k	dp	n
» VII	p	npk	dn	dk
» VIII	np	d	dnk	pk

Fig. 7.3: Plan of a 2^4 experiment on beans, partially confounded in randomized blocks of four plots.
Dung: None, or 10 tons per acre (d)
Nitro-chalk: None, or 0.4 cwt. N per acre (n)
Superphosphate: None, or 0.6 cwt. P_2O_5 per acre (p)
Muriate of potash: None, or 1.0 cwt. K_2O per acre (k)
Plot size: 1/80 acre (45 ft. × 12 ft.)
Confounding: PK, DNP, & DNK in Blocks I—IV;
DP, DNK, NPK in Blocks V—VIII.

difficult. Fig. 7.2 illustrates a 2^3 experiment confounded in 6 blocks of 4 plots.

In field experiments, confounding of 2^3 designs is rarely necessary, because a block of 8 plots is not unreasonably large; in an animal experiment, however, the convenient block might easily be smaller than 8 and the scheme just described could then be valuable. Confounding is none the less important in field experimentation, because often a larger number of factors is to be included. For example, a 2^4 experiment can be arranged in blocks of 8 by dividing the 16 treatments between two types of block and confounding the four-factor interaction. A 2^5 experiment can be arranged in blocks of 16, by dividing the 32 treatments between two types of block and confounding the five-factor interaction; it can also be arranged in blocks of 8 by dividing into four types of block and confounding a four-factor and two three-factor interactions. A 3^3 experiment can be arranged in blocks of 9 by confounding part of the three-factor interaction.

In the confounding arrangements discussed above, the whole of the potential information on certain interactions is sacrificed in order that smaller blocks may be used. If the experimenter wishes to avoid this,

I		II		III	
n_2	p	n_1p	(1)	pm	n_2pm
m	n_1	n_2p	pm	n_1m	n_2
n_2pm	n_1pm	n_1m	n_2m	(1)	n_1p
n_1	n_2m	n_1m	m	n_1	n_1pm
(1)	n_1pm	p	n_2	m	n_2p
pm	n_2p	n_2pm	n_1p	n_2m	p
IV		V		VI	

Fig. 7.4: Plan of a 3×2^2 experiment on potatoes, partially confounded in randomized blocks of six plots.

Artificial nitrogen: None, or 0.6 cwt. N per acre as sulphate of ammonia (n_1) or ammonium bicarbonate (n_2)
Artificial phosphate: None, or 0.6 cwt. P_2O_5 per acre as superphosphate (p)
Poultry manure: None, or at rate of 0.6 cwt. N per acre (m)
Plot size: 1/65 acre (30 ft. $\times$ 22 ft.)
Confounding: PM, NPM (partially).

he may distribute his sacrifices more widely by adopting *partial confounding,* so that different interactions are confounded in different groups of blocks. Fig. 7.3 shows a simple scheme of partial confounding for a 2^4 experiment in 8 blocks of 4, three interactions being completely confounded in each half of the design. With factors at more than three levels, partial confounding can be considerably more complicated, as perhaps Figs. 7.4 and 7.5 illustrate too well; the reader will need to examine these carefully before he can understand the symmetry or partial symmetry in their structure.

The number of ways in which a factorial experiment can be confounded may be great, especially if many factors are involved. Ingenuity will then be required in order to choose the form of confounding most suitable for a particular purpose, but the formation of the several types of block is a mechanical process*.

* I have given an elementary account of this in "The construction of confounding arrangements" (1947, *Empire Journal of Experimental Agriculture,* Vol. 15, 107—112).

d r	ds_2	ds_2 rc	rc	s_1 r	s_1 c	ds_1 rc	d c	s_2 r	ds_1	s_2 c	(1)

Block I

r	ds_1 c	s_1	d rc	s_1 rc	s_2 rc	ds_2 r	s_2	d	ds_2 c	ds_1 r	c

Block II

ds_2	d	s_2 r	s_1 rc	d rc	ds_2 rc	c	ds_1 r	r	ds_1 c	s_2 c	s_1

Block III

ds_1	(1)	s_2 rc	ds_1 rc	ds_2 r	s_2	d r	rc	d c	s_1 r	s_1 c	ds_2 c

Block IV

Fig. 7.5: Plan of a 3×2^3 experiment on sugar beet, partially confounded in randomized blocks of twelve plots.

Dung: None, or 10 tons per acre before winter ploughing (d)
Salt: None, or 4 cwt. per acre before winter ploughing (s_1) or before sowing (s_2)
Rolling: Ordinary, or ordinary plus additional (r)
Cultivation: Normal, or normal plus intensive inter-row hoeing (c)
Plot size: 1/40 acre (12 ft. $\times$ 93 ft.)
Confounding: DSRC (partially).

Randomization

All types of confounding are basically schemes for determining the treatments to be included in a block, these being a selection from the whole set of treatment combinations possible. Within a block, the allocation of treatments to plots must be fully randomized, exactly as for an ordinary randomized block design. Consequently, we may legitimately speak of a design as being 'confounded in randomized blocks'.

†*Statistical Analysis*

In the analysis of variance for a 2^3 factorial experiment, there will be a sum of squares with 7 d. f. for differences between the 8 treatments. Examination of the algebra that underlies the analysis of variance shows that this may be divided into 7 separate components which add to the total. Each component has 1 d. f.; one corresponds to each of the three main effects of the factors, one to each of the three two-factor interactions, and one to the three-factor interaction. Thus the significance of each component may be examined independ-

Table 7.1: Outline of analysis of variance for Fig. 7.2.

Adjustment for mean			Mean square
Nature of variation	d. f.	Sum of squares	
Between blocks	5		
Main effect of N	1		
Main effect of P	1		
Main effect of K	1		
Interaction N×P	1		
Interaction N×K	1		
Interaction P×K	1		
Error	12		
Total	23		

ently of the others. Similarly, for a 3^2 experiment, the 8 d. f. between the 9 treatments may be divided into 2 d. f. for the main effect of each factor (since there are 3 levels) and 4 d.f. (= 2×2) for the two-factor interaction; three corresponding sums of squares may be separated from the sum of squares between treatments, and these may be broken down further into single degrees of freedom for special aspects of the effects and interactions.

When confounding is introduced into a factorial experiment, the component of the sum of squares corresponding to a confounded interaction will be included in the component for 'blocks', and must not appear again under an alias as the interaction. If the confounding is complete, no entry for the interaction will appear in the analysis; if the confounding is partial, a modified calculation will be required. The further modifications needed for the special types of confounding described later in this chapter will not be discussed here.

Though this sketch of the analysis of factorial experiments is much too brief to make the reader able to analyse one for himself, it should indicate the character of the analysis. Moreover, the structure of the analysis of variance for a factorial design illustrates well the meaning of what are so often a mystery to the novice in statistics, namely 'degrees of freedom': any sum of squares of deviations can be divided into as many independent components (each with 1 d. f.) as it has deegrees of freedom, this can be done in an unlimited number of different ways, and the analysis of variance is a procedure for making this division in a meaningful manner so as to group together degrees of freedom of like character. Table 7.1 shows the outline of an analysis of variance for the 2^3 design (Fig. 7.2) with the three-factor interaction confounded.

Split Plots

One form of confounding often adopted in agricultural experiments, because it simplifies the application of certain treatments, is that known as *split plots*. Certain treatments, such as types of cultivation or applications of bulky organic manures, may not be practicable on the small areas proposed for the ultimate plots. Consequently, these treatments may be assigned to larger areas, or *main plots*, which are then split into *sub-plots* for comparisons of other treatments such as varieties of seed or applications of concentrated artificial fertilizers. The result is that the effects of the first group of treatments are confounded between main plots, though they may be estimated and their significance tested by analysis of yields for the whole of each main plot; the effects of the second group of treatments and the interactions between the two are unconfounded, and will often be estimated much more precisely. Figs. 7.6—7.8 illustrate the appearance of split-plot designs, the first two being very simple and the third more complicated

Block I			Block II		
k_2m k_2	m (1)	k_1m k_1	k_2 k_2m	k_1m k_1	(1) m
m (1)	k_2m k_2	k_1 k_1m	k_1m k_1	(1) m	k_2 k_2m
Block III			Block IV		

Fig. 7.6: Plan of a 3×2 experiment on potatoes, in randomized blocks with split plots.

Potash: None, 1.5 cwt. K_2O per acre as sulphate of potash (k_1), or mineral potash (k_2)

Magnesium: None, or sulphate of magnesia providing magnesium equivalent to 1.5 cwt. K_2O per acre (m).

Plot size: 1/60 acre (main plot).

The three potash treatments are arranged in a randomized block design, and a random half of each plot received sulphate of magnesia; the boundaries of sub-plots are not shown.

because main plots carry the combinations of two factors. All these experiments have two sub-plots per main plot, perhaps the commonest scheme of splitting, but designs with three or more sub-plots per main plot are also used.

Split-plot designs should not be adopted unthinkingly. Considerations of the practicability or convenience of an experiment may force their use, but often they involve a sacrifice of important features of the factorial scheme. The experimenter may find himself obliged to place on main plots the very treatments in which he is most interested, and must then resign himself to achieving less precision in estimating their effects than if he could avoid the splitting of plots. Sometimes, indeed, a split-plot design may happen to be ideal from the statistical point of view also, because it concentrates precision on the questions of chief interest. Another important application is for inserting an additional factor into an experiment that has already begun. A split-plot experiment has two standard deviations, one for main plot comparisons and one for sub-plot comparisons; in general the first will be larger, the second smaller than if a corresponding design without splitting had been used. Thus in Fig. 7.6 precision has probably been lost on the main effect of the potassic fertilizers in order to gain a smaller S.D. for use in testing the main effect of magnesia and its interactions with potash. This may have been the intent of the experi-

t_2 t_2n	t_1n t_1	n (1)	t_3n t_3
(1) n	t_3 t_3n	t_2 t_2n	t_1n t_1
t_3 t_3n	(1) n	t_1n t_1	t_2 t_2n
t_1n t_1	t_2 t_2n	t_3n t_3	(1) n

Fig. 7.7: Plan of a 4×2 experiment on a forage crop, in a Latin square with split plots.

Times of cutting: 11 June (no symbol), 1 July (t_1), 22 July (t_2), 12 August (t_3)
Sulphate of ammonia: None, or 0.3 cwt. N per acre (n)
Plot size: 1/80 acre (33 ft. × 16½ ft: main plot)
The four times are arranged on main plots in a 4×4 Latin square, and a random half of each plot received sulphate of ammonia; the boundaries of sub-plots are not shown.

menter, but had he been equally interested in all effects of his factors he would have done better to take the sub-plot as the basic unit and have four simple randomized blocks of six plots. In Fig. 7.7, the splitting may have been forced by the difficulty of cutting small areas of the crop at four different times. In Fig. 7.8, it is hard to see why a split-plot design was used, unless the experimenter was especially interested in the effect of phosphate and its interactions with other fertilizer effects; if this was not so, he would probably have been better advised to make the sub-plot his unit and either to have four randomized blocks of 18 plots or to adopt some scheme of partial confounding with blocks of 6 plots. The reader should at least take warning against assuming that, because a split plot design is convenient

Block I			Block II		
(1)	n_2k_1p	k_1	k_2	n_1p	n_1k_1
p	n_2k_1	k_1p	k_2p	n_1	n_1k_1p
n_1k_2	n_2p	n_1p	p	n_1k_2	n_2k_2
n_1k_2p	n_2	n_1	(1)	n_1k_2p	n_2k_2p
k_2p	n_2k_2	n_1k_1p	k_1p	n_2	n_2k_1p
k_2	n_2k_2p	n_1k_1	k_1	n_2p	n_2k_1

Block III			Block IV		
n_1k_2	n_1k_1	k_2p	n_2k_1p	n_2k_2p	p
n_1k_2p	n_1k_1p	k_2	n_2k_1	n_2k_2	(1)
n_2k_1	p	n_2	n_2p	n_1k_1p	n_1
n_2k_1p	(1)	n_2p	n_2	n_1k_1	n_1p
k_1	n_2k_2	n_1p	n_1k_2p	k_1	k_2
k_1p	n_2k_2p	n_1	n_1k_2	k_1p	k_2p

Fig. 7.8: Plan of a $3^2 \times 2$ experiment on potatoes, in randomized blocks with split plots.

Sulphate of ammonia: None, 0.3 (n_1), or 0.6 (n_2) cwt. N per acre
Sulphate of potash: None, 0.5 (k_1), or 1.0 (k_2) cwt. K_2O per acre
Superphosphate: None, or 0.5 cwt. P_2O_5 per acre (p)
Plot size: 1/100 acre (sub-plot)

Four randomized blocks of 9 main plots contain combinations of nitrogen and potash. Each plot was split in half, and phosphate assigned to one of the two sub-plots chosen at random; the boundaries of sub-plots are not shown.

for a certain purpose, therefore it is the most efficient choice: it may be, but also it may be either a compromise or even a bad choice.

†*Single Replications*

When several factors are included in an experiment, the high order interactions (that is to say, interactions of three or more factors) are likely to be small and of negligible importance; their mean squares will then differ very little from the error mean square. This property is often used to advantage by designing the experiment to have only one complete replication of the treatment combinations, so that all the degrees of freedom can be assigned to block or treatment differences, and then using a mean square from high order interactions in place of the error mean square. Of course, the experimenter must have reasonable assurance that, in the circumstances of his experiment, the interactions concerned will not be large. An example of a 2^5 design confounded into blocks of 8 plots and performed in a single replicate of 4 blocks is shown in Fig. 7.9.

One of the most valuable of single replicate designs is that for a 3^3 scheme in 3 blocks of 9 plots, confounding 2 d.f. from the three-factor interaction; an example of this is shown in Fig. 7.10. It is perhaps the most compact design available to the experimenter, and has been

Block I	nk	(1)	p	npk	Block II
	snp	sdn	sdnp	sn	
	dp	spk	d	sk	
	sdk	dnpk	dnk	sdkp	
Block III	s	sdp	n	k	Block IV
	snk	dk	sdnk	dpk	
	np	sdnpk	sp	dnp	
	dn	pk	sd	snpk	

Fig. 7.9: Plan of a 2^3 experiment on beans in a single replicate of 4 randomized blocks.

Spacing: Rows at 18 inches or at 24 inches, the latter indicated by s
Dung: None, or 10 tons per acre (d)
Nitro-chalk: None, or 0.4 cwt. N per acre (n)
Superphosphate: None, or 0.6 cwt. P_2O_5 per acre (p)
Muriate of potash: None, or 1.0 cwt. K_2O per acre (k)
Plot size: 1/40 acre (45 ft. × 24 ft.)
Confounding: SDP, SNK, DNPK.

found of great help in exploring the manurial requirements of crops. Also important are designs for 3^4 in 9 blocks of 9 plots, 2^6 in 8 blocks of 8 plots, and so on.

Block I			Block II		
$d_2n_2k_1$	$d_1n_2k_2$	d_1n_1	$d_2n_2k_2$	d_2k_1	(1)
d_1k_1	$d_2n_1k_2$	d_2	n_2k_1	d_2n_1	n_1k_2
k_2	n_2	n_1k_1	d_1k_2	d_1n_2	$d_1n_1k_1$

Block III		
k_1	d_2n_2	d_2k_2
n_1	$d_1n_2k_1$	d_1
n_2k_2	$d_1n_1k_2$	$d_2n_1k_1$

Fig. 7.10: Plan of a 3^3 experiment on beans, in a single replicate of 3 randomized blocks.

Dung: None, 7.5 (d_1), or 15 (d_2) tons per acre
Nitro-chalk: None, 0.2 (n_1), or 0.4 (n_2) cwt. N per acre
Muriate of potash: None, 0.5 (k_1), or 1.0 (k_2) cwt. K_2O per acre
Plot size: 1/60 acre (34 ft. $\times$ 21 ft.)
Confounding: DNK (partially).

Note that in Block I the suffices attached to the symbols for each plot always add to 2 or 5 (for $d_2n_2k_1$, $2 + 2 + 1 = 5$), for Block II the total is always 0, 3, or 6, and for Block III the total is always 1 or 4; this property is closely related to the confounding.

†*Fractional Replication*

With still larger numbers of factors, *fractional replication* may sometimes be profitably employed. This device involves confounding certain effects and interactions with other effects or interactions (and possibly also with blocks). In consequence, each apparent effect may have two or more *aliases:* the numerical estimate of the main effect of one factor or of an interaction may simultaneously be the estimate of one or more other effects or interactions. Such confusion might be thought fatal to the purpose of an experiment, but by careful choice of the alias associations its ill results may often be made few. For example, if each main effect has as its alias the interaction between four other factors, the experimenter will usually have little hesitation in attributing the cause of a significant result to the main effect rather than to the interaction. Fractional replication should be used with great care, and ought not to be adopted without the advice of a statistician, but it can be a powerful aid to the extensive exploration of a large number of factors. Thus a one-third replicate of a 3^5 scheme can be arranged in 9 blocks of 9 plots so as to permit the extraction of almost complete information on main effects and two-factor interactions; this design has been found valuable in exploratory studies of the manurial requirements of tropical crops, and the reader will scarcely need persuading of the advantages of being able to study 243 treatments combinations with only 81 plots! Similarly, a one-half replicate of a 2^7 or a one-quarter replicate of a 2^8 scheme can be put on to 64 plots. The greater the number of factors, the smaller is the fraction that need be used. Fractional replication should seldom be adopted with less than five factors.

Long-term Experiments

In the publication mentioned earlier, Crowther said:

> "It is assumed that the annual experiments will be repeated for several years with new sites each season. It is unsafe to judge fertilizer responses from one or two years only. The writer recently had a series of sugar beet experiments which showed no response to sulphate of ammonia in two consecutive seasons, though they had given clear responses in the 10 preceding years. Quite misleading conclusions might have been drawn if the experiments had been limited to these abnormal seasons."
>
> "At experiment stations it would be sound policy to concentrate the field work on experiments planned to run over a term of years or else requiring operations and observations too difficult or detailed to be made elsewhere. Single year tests can generally be made at a series of commercial farms or out-stations."

The need for continuation of experiments over several years is of course as great for tests of varieties or of cultivation and management treatments as it is for fertilizer tests. Moreover, with an annual crop, the growing of a second crop in the next year on land that has been used for an experiment may give valuable information on the residual effects of treatments.

These long-term programmes do not introduce novel statistical difficulties. Experiments in which crop-rotations form part of the subject of study, or experiments on crops that are not replanted annually (such as pasture, fruit, sugar-cane, and tree-crops) tend to be more complex. One important principle now generally recognized is that the different phases of a rotation experiment should be tested every year. For example, in a study of the manuring of a standard four-course rotation, each of the four crops should be represented every year by a full set of plots for the manurial treatments. Only so can the effects of seasonal differences be eliminated.

Perhaps the most important point to emphasize for any reader of this book who is planning a long-term programme of experimentation is that he should consult a statistician well before he proposes to begin. Even a year in advance of the start of the experiment is not an excessive margin, for consideration of the relative merits of different designs takes a long time and is better done in short spells over a period than in a single concentrated effort.

Animal Experiments

Although factorial design has been discussed primarily in terms of field plot trials, the same ideas can be used in experiments on animals. For example, an experiment on pig feeding at Rothamsted Experimental Station in 1935 had a 2^4 design. The factors were minimal *v.* liberal allowance of green food, unrestricted *v.* limited feeding, coarse *v.* fine grinding of food, and exercise *v.* no exercise. Eight litters of 8 pigs were used, the four-factor interaction being confounded between litters. Thus, as commonly happens, litters played the part of blocks, a device which had the further advantage of allowing four sections of the experiment (each a pair of blocks) to begin at different times. In addition, certain interactions had to be partially confounded with sex within litters, according to a scheme that will not be described here.

CHAPTER 8

SAMPLING

"A very intelligent gentleman assures me (and he speaks from upwards of forty years' experience), that the mean rain of any place cannot be ascertained till a person has measured it for a very long period. 'If I had only measured the rain', says he, 'for the four first years, from 1740 to 1743, I should have said that the mean rain at Lyndon was 16½ inches for the year; if from 1740 to 1750, 18½ inches. The mean rain before 1763 was 20¼ inches, from 1763 and since 25½ inches, from 1770 to 1780, 26 inches. If only 1773, 1774, and 1775 had been measured, Lyndon mean rain would have been called 32 inches'."

Gilbert White: "The Natural History of Selborne".

The Purpose of Sampling

We have already considered (Chapter 5) one simple example of a sampling procedure. We must now discuss sampling problems in greater detail.

Suppose that a certain property of a *population* or *universe* is to be measured*. Economy of time or effort, or other important reasons, may demand that only a *sample*, a portion of the population selected according to appropriate rules, be measured, and that the result obtained be then assumed to apply approximately to the whole. An obvious instance is that for which the process of measurement destroys

---

* The statistician uses the words population and universe to describe any assembly of material (not necessarily 'large') from which he proposes to discuss selecting a part. Examples are:— the inhabitants of Great Britain on a particular day, all fields of potatoes in a country, all cows in milk at any time during 1952 on a particular farm, a bag of seeds, all farms over 500 acres in England, a particular field of wheat (from the point of view discussed later in this chapter), today's milk from one cow, and the roots of a single plant (from which perhaps a part may be taken for chemical analysis). Even a purely abstract or conceptual assembly may be a statistical population, such as all possible 5×5 Latin squares (from which one is to be selected for an experiment), all possible results from throwing three dice simultaneously, or all possible numbers of cockerels that might be hatched from eleven eggs.

the individual measured. A seed merchant may wish to quote in his catalogue the percentage germination for seeds of various species, but he can determine whether or not an individual seed will germinate only by direct test and thus by rendering it unfit for sale. If he wishes to escape from the dilemma of either selling without knowledge of germination or testing all his supply and so selling none, he must rely on a sample: he must select a number of seeds, with precautions to ensure that those selected are reasonably representative of his stock, test these for germination, and assume that the bulk of his stock of this species has approximately the same percentage germination.

The reader might at first imagine that reliance upon a sample was necessarily unsatisfactory, and was to be resorted to only when (as in the example just quoted) complete measurement is impossible or prohibitively laborious. In practice, however, a well planned sample may often be preferred to complete measurement, because it can be so chosen as to give results sufficiently precise for the use the investigator has in mind yet the data can be obtained and analysed much more speedily or economically than can those for complete measurement. Moreover, if only a relatively small sample of a large population has to be studied, the persons responsible for making the measurements and observations can often be of a higher standard of ability than if the whole population were included. This point is particularly important in extensive survey work, such as a survey of fertilizer practice. If information on fertilizer applications had to be obtained for every farm in a county, either the questionnaire would have to be kept very simple or a large corps of assistants with no great knowledge of, or interest in, the problem would have to be employed; concentration of effort on a small number of farms brings the work within the compass of the small number of skilled persons available, and enables much more detailed information to be sought*.

A sample is some part of a population, selected from the whole in accordance with rules that have been carefully specified. Measurements or observations are made on the sample (or on its individual

---

* A paper entitled "The manuring of farm crops: some results of a survey of fertilizer practice in England", by F. Yates, D. A. Boyd, and I. Mathison (1944, *Empire Journal of Experimental Agriculture,* Vol. 12, 163—176) gives a clear account of the valuable information obtained in such a survey. Resources were limited, but expert staff and a carefully planned scheme of sampling enabled these to be used to best advantage. Anyone who reads this paper will realize that, had a much larger number of farms been included, the standard of interviewing of farmers would have suffered and consequently the trustworthiness of data from individual farms would have been reduced.

members) with a view to calculating from them a value that shall estimate some characteristic of the population. For example, members of a sample of the population of Great Britain might be asked questions relating to their income or to their weekly consumption of beer, with a view to estimating the frequencies of different income levels or of different beer-drinking levels in the population; a sample from the fields of potatoes in a county might be selected and measurements of yield or of blight infection made on its members in order to estimate the total yield or the mean percentage blight for the county; a daily sample from the milk of one cow might be analysed in order to estimate the percentage butter-fat in her whole yield for the day. Often one sample is used for many different but related purposes. For example, in the survey of fertilizer practice already mentioned, records of certain fields on a sample of farms within a county were used to give information on mean rates of application of each of several types of fertilizer in the year of survey and also in previous years. The ratio of the size of the sample to the size of the whole population is termed the *sampling fraction;* thus, if a sample from 660 potato fields in a county consists of 30 fields, the sampling fraction is 1/22.

Sampling and the Statistician

The contribution of the statistician to the practice of sampling is as important as, and is analogous to, his contribution to experimentation. Not only will he analyse the data, but he will also advise on the design of the sample. His analytical function is obvious: he will be responsible for reducing the data so as to obtain estimates of the quantities that interest the experimenter, and for expressing the precision of these estimates by the use of standard errors, fiducial limits, or other devices. An estimate is generally of little use unless accompanied by some indication of the error to which it may be subject. Of course, if the exact amount of any error were known, the estimate could be corrected by a simple adjustment, but in practice all that can be done is to determine a standard error and from it find limits between which the true value is almost (but not absolutely) certain to lie. An example of the appropriate form of statement is given at the end of the next section.

We have seen earlier that an experimenter, anxious to make the most precise comparisons between various treatments that his resources will allow, can be helped by the statistician who will suggest a suitable experimental design. In sampling also, the statistician who has access to information on previous similar investigations will be able to propose a design for the sample — in other words, a set of rules by which the sample is to be selected from the population and measured — likely to lead to the most precise estimate for a given expenditure of time and materials. Indeed, the need for expert advice on the economic allocation of resources is especially evident in connexion with sampling. The mean yield per acre of potatoes in a county could be determined almost exactly by weighing the whole produce of every field, but the labour required for this would be prohibitive. Much more important in practice is consideration of the best employment of a limited amount of effort: should the whole produce of a number of fields be measured, and, if so, should these be concentrated on a few farms or spread over many? Or should only the crop from small areas on a larger number of fields be weighed? How are the fields and the areas within fields to be selected from the potato-growing land of the county? Is it practicable to devise a scheme that will give an estimate of the mean yield of the county almost certainly within 5 cwt. per acre of the truth, and what is the most economical way of doing this? These questions the statistician may answer, at least in part; he cannot hope to be perfectly right, because he must base his advice on past work and the current investigation is sure to differ in some respects. Moreover, if one sample is to be used for estimating several different quantities, its design must be a compromise between the designs that would apparently be the most economic for each separately.

However strongly the statistician urges that he be consulted before a sample is taken rather than after the data have been collected, inevitably he is sometimes asked to advise on, or to assist with, the analysis of a sample whose collection he did not design. To do this, he must know not only the numerical values constituting the data but also the way in which these were collected: in fact, although he did not design the sample, he must know the design that was used. Just as for an experiment, the character of the design determines the appropriate form of analysis. Unless he is told exactly how a sample was selected, the statistician cannot safely form an estimate or assess precision: indeed, he cannot even be sure that either of these calculations is possible, since a thoroughly bad design might prohibit all legitimate estimation.

Sampling for Yield Estimation

One very simple sampling problem was discussed briefly in Chapter 4. We now turn to another problem, of a type that is often encountered, and discuss it in greater detail. Not only does it illustrate well the calculations required in the analysis of a sample, but it also gives scope for an elementary discussion of the design of a sample.

Suppose that the mean yield per acre of a field of wheat is wanted, and that the inquirer is either unable to measure the yield for the whole field or unwilling to wait until the completion of the harvest makes this figure available. Instead, small areas of the crop may be cut by hand, the grain may be threshed out from these, and the yield per acre estimated by the yield per unit area for the sample. We shall later consider the question of deciding the best size for these small areas or *sampling units**, but for the present we may discuss the problem in terms of units 1 yard square (about the order of magnitude generally found advisable for a cereal crop). These units will all be placed with one side parallel to the direction of drilling, and, in order to ensure that the same number of rows is always included, their width across the drills should be a multiple of the distance between rows. Thus, if the crop were sown at 6 inch spacing, this unit would contain 1 yard of six adjacent rows; if the spacing were 8 inch, a unit 1 yard × 32 inches would be preferred, so as to give 1 yard of four adjacent rows.

If an investigator were to go into the field and himself select 'haphazardly' a number of places from which to cut sampling units, he would almost certainly produce a biased estimate. Despite all his efforts to be fair, he might tend to pick areas where the crop was taller and heavier, or, in a conscientious attempt to avoid this bias, he might go to the opposite extreme. Experimental trials of many types of sampling have demonstrated how easily bias may enter if the sampler is permitted to exercise subjective judgement. Instead, he should ensure that every part of the field has the same chance of inclusion in the sample** and should then introduce some form of ran-

---

* The distinction between 'sampling unit' and 'sample' is important if confusion is to be avoided, but many publications fail to make it clear. The sample is the totality of sampling units from which an estimate is to be made, and the word should not be applied to the units separately. Sometimes measurements have to be made on distinct parts of every sampling unit, as when the unit consists of several plants each of which contributes a measurement for the records: these parts of the sampling unit may be termed *recording units.*

** In stratified sampling (see below), the sampling fraction is sometimes deliberately increased for strata in which the variability is large, since this may be shown to improve the precision of the final estimate. This practice does not really conflict with the principle enunciated here, since the essential point is equality of chance of inclusion for every possible sampling unit within a stratum; the statistical calculations for stratified sampling, however, must be modified when a variable sampling fraction is employed.

dom selection in the placing of the units. The simplest scheme to adopt is that of *unrestricted random sampling* (analagous to the completely randomized design in Chapter 5): an agreed number of sampling units are taken at points selected entirely at random within the field. This is easy to do if the field is rectangular. The sampler first

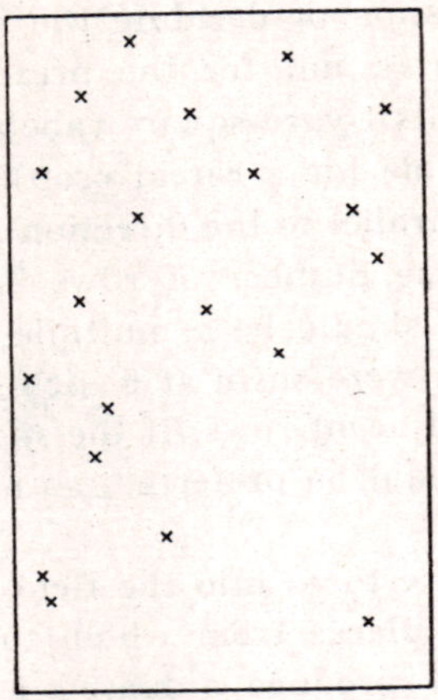

Fig. 8.1: Unrestricted random sample.
Positions of 20 sampling units in a rectangular field.

measures the length and breadth, in terms of his normal pace-length if he wishes to avoid using a tape; suppose the dimensions to be 300 yards along the rows, 170 yards across the rows. He next chooses a number at random between 1 and 300 (say 43), and a second at random between 1 and 170 (say 119); he may use numbered cards (as described in Chapter 2) or published "Tables of Random Numbers" for this. Taking with him a light wooden frame 1 yard square, he walks 119 paces along a short side from one corner, turns into the field and walks 43 paces along the rows. He puts down the frame on his left, with two sides parallel to the drills and the front right-hand corner as close as possible beside the toe of his leading foot. He then cuts the stems within the frame, taking care to ensure that he has the right number of rows included and that stems near the edge of the frame are correctly treated. All this he repeats with other pairs of sampling numbers, until he has as many units as he wants; Fig. 8.1 shows how a set of 20 units might be situated in a field. The investigator will usually prefer to select all his sampling numbers before he begins any cutting. The produce of each unit is kept separately for subsequent measurement of yield. If the field is less regular in shape or contains obstructions such as trees and ponds, the advice of a statistician on the selection of random points will be needed. The reader will appreciate that the careful instructions given here are intended to eliminate all subjective bias; many of the details may be varied, provided that the specification is kept objective and fairly rigid.

There are undoubtedly occasions when strict random sampling of this kind is impracticable. The reader must be on his guard, however, against uncritical adoption of some 'haphazard' scheme that appears to him to have all the merits of random sampling and much less of its labour. In pasture sampling, for example, a walk over the field with an occasional stop for flinging a stone and taking a sampling unit where it falls has been suggested; this is tempting in its simplicity, but probably involves a subconscious avoidance of the margins of the field, of the neighbourhood of trees, of damp patches, and so on, all of which may bias the estimation. If any deviation from true randomness is contemplated, its consequences should be most carefully considered and a statistician should be asked to say what disadvantages may accompany the readily apparent advantages.

We now suppose that the sampling has been performed, that 35 sampling units were taken, and that the weights of grain from these (in ounces) were:—

6,	13,	8,	7,	10,	8,	9,	10,	11,	8,	8,	9,
7,	9,	7,	6,	6,	7,	7,	8,	9,	5,	3,	6,
10,	8,	8,	7,	9,	7,	8,	5,	9,	9,	8	

The total weight of grain from these 35 square yards is 275 oz., and the mean 7.9 oz. Since 1 oz. per square yard corresponds to 4840/(16×112) or 2.70 cwt. per acre, this weight represents 21.3 cwt. per acre. Hence, if the sampling units can be regarded as typical of the whole field, 21.3 cwt. per acre is an estimate of yield for the field.

The standard deviation per sampling unit may be calculated exactly as in Chapter 5. The sum of squares of deviations from the mean is

$$^2 + 13^2 + 8^2 + 7^2 + \ldots\ldots\ldots\ldots + 9^2 + 8^2 - \frac{(275)^2}{35} = 118.3.$$

This has 34 degrees of freedom (35 units); therefore the estimate of variance is

$$s^2 = \frac{118.3}{34}$$

$$= 3.48,$$

and the estimated S.D. is

$$s^2 = \sqrt{3.48} = 1.87.$$

This is measured in ounces, the same units as the original records. The standard error of the mean of 35 units, by the rule in Chapter 4, is the S.D. divided by $\sqrt{35}$, and is therefore 0.316 oz. Multiplication by 2.70 converts this to 0.85 cwt. per acre. The information may be summarized in the statement that the yield of the field is estimated to be 21.3 ± 0.85 cwt. per acre. This may be interpreted in the language of fiducial limits, as explained in Chapter 5. For example, the value of t for 34 degrees of freedom at the 0.05 probability level is 2.03, so that a deviation from the true mean exceeding 2.03×0.85 (or 1.7) cwt. per acre has a probability of only 0.05. Hence the investigator may reasonably assert that the true yield lies between 19.6 and 23.0 cwt. per acre; such an assertion may be false in any one instance, but if he habitually draws inferences in this way he will be correct in about 95 % of his statements.

Size of Sampling Unit

In the last section, the investigator was assumed to have chosen a sampling unit 1 yard square. We now need to discuss how such a choice should be made. By taking a large enough number of units of this size (or of any other size), randomly situated in the field, he could make the standard error of the yield estimate as small as he wished, but he will naturally prefer to attain a desired precision by choice of the most economical unit rather than by prodigal use of an arbitrarily defined one. A unit much smaller than 1 foot square might be impracticable because slight inexactness in the determination of the boundaries could have a relatively large influence on the stems cut for the unit. A unit much larger than 10 yards square would probably be quite impracticable for the small-scale threshing and other equipment usually employed, although this difficulty might be overcome if a large unit were found to have other advantages. Between these extremes, a wide choice of shapes and sizes confronts the investigator. He cannot hope to make the best possible choice, because no two fields are alike and only intensive study of the very field he proposes to sample could give him the information he needs for that. Nevertheless he may either wish to be guided by the experience of previous samplings or plan to use his current study to make possible a better choice on the next occasion.

On general grounds, we may except that the information provided by one sampling unit will be smaller the smaller the size of the unit: the S.E. of a mean based on 35 units each of 1 foot square would be greater than for an equal number of units 1 yard square. This is obvious, because the estimate is being based on a smaller proportion of the crop, with no advantages to compensate. At the same time, the expenditure of time and effort on sampling is likely to decrease with the size of unit. In wheat sampling, a small unit gives less produce to be carried than does a large, and the threshing will take less labour. The decrease will not be in proportion to the size of unit, because certain factors are almost independent of size (time spent in choosing and locating sampling points, in recording data, and in statistical analysis). On the other hand, for a fixed total size of sample, division into many small units is likely to give greater precision than division into a few large units: the S.E. of a mean based on 35 units 1 yard square would be smaller than that of a mean based on 5 units measuring 2 yards×3½ yards, because the smaller unit permits a wider distribution of sampling points over the field. This suggests that, if economic considerations were governed entirely by the total size of the sample (that is to say, the total area within all the units), the smallest permissible sampling unit would be the best. In practice, other difficulties may put the smallest permissible unit at an economic disadvantage relative to one rather larger, in spite of the larger S.E.; in particular, the separate cutting, processing, and recording of a very large number of small units may be so inconvenient as to make preferable a slightly larger total sample size provided that the latter is spread over fewer units. At this stage, appeal to observation and to knowledge of relative costs is necessary before more can be said about the practical merits of different sizes of sampling unit.

In sampling a crop consisting of larger individual plants, such as sugar beet or potatoes, definition of the sampling unit in terms of a number of plants (e. g. 8 successive plants in a row, or two runs of 4 plants in adjacent rows) may be preferred. Random location of the unit is still essential. If the purpose of the sampling is yield estimation (rather than, say, study of the incidence of disease), the area or length of row occupied by each sampling unit should be measured.

Shape of Sampling Unit

In the sampling of crops such as wheat, square or rectangular sampling units are usually preferred. Circular units have the advantage of being easily marked out with a central peg and a radial string of fixed length; triangular units are easily marked with three pegs joined by strings to give the sides. Both these shapes have been advocated as avoiding the necessity for carrying a rigid frame. In a crop sown broadcast, this might be a gain worth exploiting; in a drilled crop, increased variability might result from the unequal lengths of drill included in different units. Squares and rectangles are likely to be found the most generally useful.

If nothing is known about variations in fertility and other factors influencing growth of the crop, a unit that is approximately square will be a better choice than an elongated rectangle of the same area. Rectangular units whose longer sides follow the direction of any major change in fertility, however, will tend to show smaller S.Ds. than squares of equal area, because they sample different conditions more widely; whether or not this effect is great enough to be important within the practicable range of dimensions is a matter for special investigation in each new set of circumstances.

Relative Efficiency of Alternative Units

We will now suppose that, in addition to the data on wheat sampling analysed in an earlier section, the investigator has obtained records from the same field with the aid of three other types of sampling unit. The unit of 1 yard square may be called type A. Unit B is 2 yards along the rows × 1 yard across (2 yards of 6 adjacent rows), and the weights of grain (in ounces) from 33 such units placed at random were:—

11, 20, 14, 14, 19, 19, 12, 18, 20, 19, 17,
17, 18, 21, 20, 18, 14, 20, 15, 14, 15, 12,
12, 14, 17, 13, 18, 14, 14, 19, 9, 14, 18.

Unit C was the same size, but was 1 yard along the rows × 2 yards across (1 yard of 12 adjacent rows). For 26 units, the weights (in ounces) were:—

19, 16, 16, 19, 12, 16, 19, 17, 16, 15, 18, 19, 13,
19, 16, 17, 13, 14, 10, 13, 11, 17, 13, 12, 13, 16.

Finally, 19 units of unit D, which was 2 yards square, had the weights (ounces):—

32, 35, 38, 34, 36, 30, 36, 36, 32, 28,
32, 34, 24, 22, 35, 29, 34, 32, 24.

Table 8.1 summarizes results for these four types of unit, the calculations being of exactly the same form as described for type A; the reader ought to verify at least one of the other columns for himself.

The first point to be noticed is that the estimates from the four types of unit agree well, as indeed they ought; all suggest that the true yield of the field is about 20—22 cwt. per acre. S.Es. of these means, or the fiducial limits, indicate the precision actually obtained, but do not necessarily show the relative merits of the four units, since

Table 8.1: Results for four different types of sampling unit used in sampling a field of wheat.

	Type of unit			
	A	B	C	D
No. of units	35	33	26	19
Weight per unit (oz.) ...	7.9	16.0	15.3	31.7
Variance	3.48	9.78	7.36	20.54
S.D. (oz.)	1.87	3.13	2.71	4.53
S.E. (oz.)	0.316	0.545	0.531	1.039
Yield (cwt./acre)	21.3 ± 0.85	21.6 ± 0.74	20.7 ± 0.72	21.4 ± 0.70
Fiducial limits (cwt./acre)	19.6 23.0	20.1 23.1	19.2 22.2	19.9 22.9

there is no reason to suppose that the total numbers of each type of unit represent equal expenditures of time or money. For a proper comparison, the S.Ds. must be used to predict the precision that would have been obtained with other numbers of units. One interesting basis of comparison is to assume that the cost of sampling is determined only by the total area of the sample. Then, if the investigator is prepared to include, say, 40 square yards in his sample, except for the question of precision he is indifferent whether he has 40 units of type A, 20 of B or C, or 10 of D. We may inquire what precision he would have obtained with each of these: on the assumption that the S.Ds. remain as in Table 8.1, the answer is given in Table 8.2. For example, the S.E. for the mean of 20 units of type B would be $3.13 \div \sqrt{20}$ oz. per 2 square yards, or 0.700 oz. per 2 square yards, which is equivalent to 0.95 cwt. per acre.

The relative magnitudes of the S.Es. in Table 8.2 are very different from those in Table 8.1. Unit A is now established as the best of the four, as would be predicted by the general argument given earlier, but it is closely followed by unit C, with B and D appreciably inferior. Though 40 square yards has been arbitrarily chosen as a total area, the same relative magnitudes would have been found whatever the choice. The result has come too late to help the sampling of this field, but the investigator may be guided by it in any future sampling of a similar nature.

These are artificial data, and no fundamental importance should be attached to the numerical conclusions, but the reader may like to note the difference between B and C, two units of the same size. Such a difference might well occur in practice when units of the same size are differently oriented in relation to the direction of drilling. Moreover, a long narrow unit often proves better than a square unit of the same area, because it cuts across patches of the crop that may vary widely in quality. To some extent, the nature of a crop limits the size and shape of units that can be used in sampling it, but, other things being equal, length across the direction of drilling is usually the ideal to aim at for a specified area of unit*.

Table 8.2: Precision obtainable from a sample containing in all 40 square yards of crop.

	Type of unit			
	A	B	C	D
No. of units	40	20	20	10
S.E. (oz.)	0.296	0.700	0.606	1.433
S.E. (cwt./acre)	0.80	0.94	0.82	0.97

A comparison based on total sample area may sometimes be a reasonable approximation. For example, in sampling for the estimation of infestations of wireworms and other soil insects, the sample will consist of small unit areas of soil dug out to a uniform depth, from which units the insects will later be extracted and counted. The greatest part of the labour is likely to be this extraction, and hence the total cost of a sample will be nearly proportional to its area, though even then certain factors such as location of sampling points and labour of statistical analysis will be more nearly proportional to the number of units than to their total area.

A better evaluation of the relative merits of different units may be made if the investigator can make a reasonable guess at the cost per unit of each type Even a crude guess is better than none. For example, realizing that he cannot do two samples of type A quite as easily as one of type B, he might guess the relative costs for A, B, C, and D to be 1.0, 1.7, 1.7, and 3.0. He can now compare any reasonable number of units of type A with the equally costly numbers of the other types:

* We have here assumed that only square and rectangular units are to be considered.

instead of 45 units of type A, he could equally well take 45 ÷ 1.7 or 26 of type B, and so on. Thus he constructs Table 8.3, the subsequent steps being exactly as for Table 8.2.

On this basis, unit C comes out as the most economical, in the sense that with it a specified expenditure would have given a smaller S.E.

Table 8.3: Precision obtainable from equally costly samples with the four types of unit.

	Type of unit			
	A	B	C	D
Relative cost	1.0	1.7	1.7	3.0
No. of units	45	26	26	15
S.E. (oz.)	0.279	0.614	0.531	1.170
S.E. (cwt./acre)	0.75	0.83	0.72	0.79

to the estimates of yield than would equal expenditure on any other unit. We must not forget, however, that the S.Ds. in Table 8.1 are themselves only estimates, and the same values would not be exactly reproduced even in further sampling from the same field. The differences in Table 8.3 are small, and certainly not statistically significant, but they give some indication that C is the best choice. No one with experience of sampling would be prepared to base decisions on policy for future work on the result of one study of this kind, but by collecting similar records from other fields a reasonable agreement might be found either that one unit was almost always the best or that several were so closely equivalent as to make the choice between them unimportant. Though comparisons of different units are made most satisfactorily on the same field or fields, work on different fields may also be brought into a discussion of relative merits.

We have discussed the choice of a sampling unit by considering the precision obtainable by equal expenditures of time, money, or labour on alternative units. We should have reached exactly the same conclusions had we adopted the alternative approach of asking which unit would allow a specified precision (i. e. a specified S.E. for the estimate) to be achieved most economically.

Other Applications

Essentially the same principle as has been used in the example discussed above finds many applications in agriculture. Indeed, for the chief crops, much work of the same character has been done, reported, and digested by statisticians; when further sampling is contemplated, a statistician can often draw on well-known publications in order to make some suggestions about the size and shape of unit. Nevertheless, knowledge is still very limited because conditions vary so widely, and no purely mathematical theory can replace the need for extensive knowledge of what does happen with various crops. Sampling is essentially a practical instrument of research, and in choosing a sampling scheme due weight must always be given to its cost and convenience; within the limits thereby set, the statistician may be asked for help in the choice of unit and of other features of design to be discussed below.

Every new application of sampling is likely to introduce its own problems. A unit that seems excellent for wheat may be at least moderately satisfactory for other cereals, but may be much less good for pasture and quite inappropriate for root crops. When individual plants are easily distinguishable, a unit based upon counting of plants (e. g. five consecutive plants randomly situated in a random row) may often be easier to use than one defined by its linear dimensions. Moreover, S.Ds. for sugar beet or potatoes will bear no relation to those for units of equal size for cereals. If the purpose of the sampling is to assess the incidence of a disease, the degree of insect attack, or the density of an insect population, the unit and the design may again need altering. In every case, a statistician may be able to help, perhaps because of his familiarity with the literature of sampling and therefore his knowlegde of the steps taken in analogous investigations in order to obtain high precision, perhaps because the problem is amenable to mathematical analysis in order that a precise sampling scheme may be devised, or perhaps because he can say with fair certainty that a standard error low enough to make the estimate worth having is quite unattainable without an impossible expenditure of time and labour.

Essentially the same methods are used in sampling fields for the study of disease and insect infestations, and also in sampling population of animals. The reader will remember that Chapter 2 contained a strong warning against using statistical methods appropriate to discrete data in the analysis of continuous measurements. He may now note that to analyse discrete data as though they were continuous is usually a less serious crime, though in many cases (such as the example in Chapter 2) either it is obviously unsuitable or it involves neglect of important information. In the estimation of the total number or the density of an insect population, however, we may proceed just as for yield estimation; we take sampling units from the insect's habitat and determine a mean and S.E. from the numbers counted in these units. The numbers constitute discrete observations, but to calculate and use S.Es. in the ordinary way is seldom misleading unless the numbers per sampling unit are very small or very variable*.

Bias, Precision, and Accuracy

The word *bias* has already been used. The reader must not take it as meaning simply the difference of the estimate from the true value. Bias is a property of the sampling procedure and not of the particular sample taken. A sampling procedure is said to be unbiased if the rules by which the sample is selected are such as to ensure that the average result obtained from many independent repetitions of the procedure (i. e. many independent samples) would be the true value for the population. For example, the mean height of undergraduates in a university might be estimated by measuring the heights of an unrestricted random sample of 50 taken from a complete list of names: if the random selection were properly performed, every possible set of 50 would have the same chance of being chosen, and the procedure would be unbiased. On the other hand, a sample defined as the first 50 to emerge from a certain college during a period of observation might easily be biased because members of that college were not typical of the whole university in respect of height. A sampling procedure for crop yield estimation that fails to give the edges of a field fair chance of inclusion will be biased, even though a particular sample might happen to give a result agreeing perfectly with the yield for the whole field: repetition of the sampling will always leave the edges under-represented, and so on an average will give a wrong value. A statistician should ensure that a sampling scheme is either unbiased or subject only to a bias that is negligible relative to the S.E.

Freedom from bias, of course, relates to the measurement of the population as it is at the moment of sampling and by the same procedure as is used for the sample. If the yield of a field of wheat is estimated by sampling a few days before harvest, the figure obtained may for various reasons be higher than a value obtained by subsequent measurement of the whole crop harvested: bad weather or attack by birds may reduce the crop between sampling and harvest, and the sample is likely to be much more carefully handled than the whole crop so that loss from shedding of grain is minimized.

By the *precision* of the estimate is meant the closeness of agreement

---

* A large investigation of this kind is fully described in *Wireworms and Food Production*, Bulletin no. 128, Ministry of Agriculture and Fisheries (1944), and in "Statistical problems in field sampling for wireworms", by F. Yates & D. J. Finney (1942, *Annals of Applied Biology*, Vol. 29, 156—167).

to be expected between a succession of independent estimates formed by repetition of the sampling procedure. In its technical sense, precision does not necessarily connote agreement between the sample and the true value unless the sampling is also unbiased*. Novices often imagine that, in order to compare the precisions attainable with various numbers of sampling units of one kind, they must make independent trials with each number. In fact, one of the important features of most good sampling methods is that, from one sample alone, the properties of samples using different numbers of units within the same scheme of random selection can be investigated: this is a consequence of the square root rule for the S.E. of a mean, explained in Chapter 4. On the other hand, comparison between different units generally necessitates independent trial of each. We have discussed examples of both processes in an earlier section of this chapter.

Some writers have used the word *accuracy* to represent a compound of precision and bias, the whole departure from the true value. In order to avoid confusion, I prefer to keep the word accuracy in its colloquial sense as relating only to correctness of measurement, of recording, and of calculation.

Stratification

In unrestricted random sampling, as explained above, sampling units are taken from random positions anywhere within the region to be sampled. By the chances of random selection, in any one instance the locations of the units may happen to be mostly in one part of the region sampled. For example, if 20 sampling units are to be located at random in a rectangular field, they may happen to be predominantly towards the northern end of the field. This would not imply that the sampling procedure was biased, because in any repetition of the sampling the distribution of the units would be just as likely to favour the southern end of the field; it would however lead to low precision, in the sense that, on account of the variability in the locations of the units, repetitions of the sampling process might give results differing very widely from one another and from the true value.

The objection to this type of sampling is like the objection to completely randomized designs for experiments (Chapter 5); the way of

---

* I have given an elementary but more extended discussion of these and other fundamental ideas about sampling in a paper "Volume estimation of standing timber by sampling" (1947, *Forestry*, Vol. 21, 179—203). Although the field of application there considered is rather different, the similarity is sufficient to make the paper helpful for readers of this book.

overcoming it is clos ly allied to the randomized block design (Chapter 6). The region to be samp ed is divided into parts or *strata,* which may be (but need not be) equal in size, the main consideration governing the choice of a stratum being that it should be as homogeneous in respect of the quality studied as the investigator can arrange. The

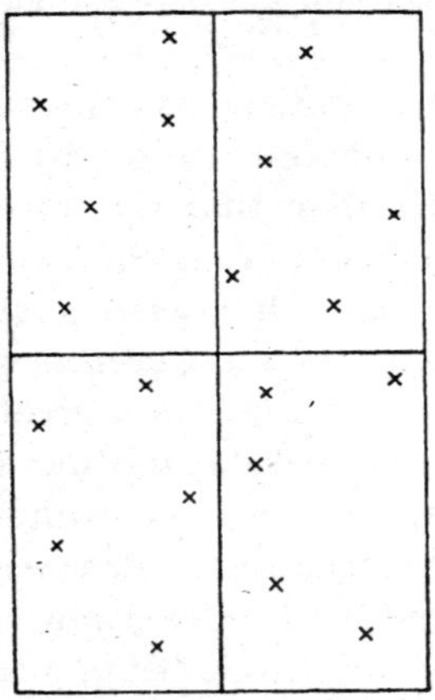

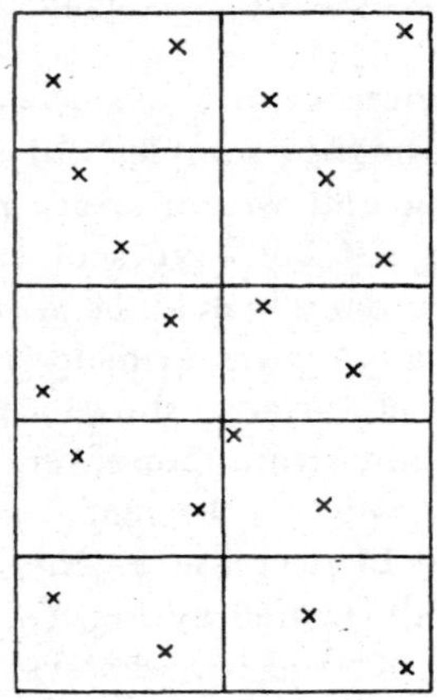

Fig. 8.2: Stratified random sample.
Positions of 20 sampling units in a rectangular field divided into four strata.

Fig. 8.3: Stratified random sample.
Positions of 20 sampling units in a rectangular field divided into ten strata.

total number of sampling units to be selected is then allocated between the strata in proportion to their size, and the appropriate number of units is taken from randomly determined points within each stratum. Instead of taking an unrestricted random sample of 20 sampling units from a rectangular field, the field might be divided into quarters and 5 units taken at random points within each quarter (Fig. 8.2), or it might be divided into tenths and 2 units taken within each (Fig. 8.3). Such a *stratified random sample* is as fair and unbiased as an unrestricted random sample in representing the field, since each part contributes to the sample in proportion to its size. At the same time, the design of the sample prevents any chance concentration of sampling units in one part of the field and insists that they be well distributed. The estimate of the mean is unaffected by average differences between strata (just as treatment comparisons in a randomized block design are unaffected by average differences between blocks), and the standard error of the mean will be calculated from a mean square within strata. An analysis of variance of the measurements on the sampling units in Fig. 8.2 can be used to divide the 19 d. f. into 3 d.f. between strata and 16 d.f. within strata, and the S.E. is calculated from the mean square for the last of these. Similarly the 19 d. f. for Fig. 8.3 can be divided into 9 d. f. between strata and 10 d. f. within strata, the last again being the basis of the assessment of the S.E.

Any increase in precision of the stratified sample, as compared with an unrestricted sample, will appear as a consequence of the S.D. per sampling unit within strata being smaller than that for unrestricted sampling. If both types of sample were taken from the same field, this difference would be directly measurable, but even a stratified sample alone gives evidence of the existence of a difference when its analysis of variance shows a mean square between strata greater than that within strata. Moreover, the smaller the strata, the more homogeneous can they be made, and therefore the variance within strata is likely to decrease as the size of each stratum is decreased: the scheme illustrated by Fig. 8.3 may be expected to give a more precise estimate of yield (or whatever property of the field is being measured) than that illustrated by Fig. 8.2. This suggests that we might with advantage go one stage further, divide the field into twentieths and take one unit at random within each. Probably the variance within strata would then be less than for the stratification in Fig. 8.3, but the sample itself would not permit estimation of this variance; the 19 d. f. between the sampling units would now all be 'between strata', and none would remain to measure the intra-stratum variation that would occur if the sampling were repeated. Although this procedure is occasionally useful, the safest general rule is that each stratum should be as small as is practicable yet large enough for each to have at least two sampling units selected from it. To use still smaller strata and to take only one sampling unit from each may give a more precise result but will deny to the investigator any possibility of assessing the precision in terms of the S.E. or the fiducial limits.

The remarks above have rather assumed that the strata are arbitrarily defined by the investigator. In some applications of sampling methods, administrative convenience may make a particular choice of strata desirable, even though it may not be the best possible. For example, in a sample survey of farms, for the estimation of a quantity such as mean rate of fertilizer application to potatoes or annual milk production, considerable variation between different parts of the country would be expected, and either counties or regions of similar farming type might be adopted as strata; to use smaller strata might add enormously to the complexity of selecting the sample, and would possibly achieve little in respect of precision once the major differences had been removed. Stratification according to size of farm might also be useful, and could be combined with a regional classification. The sampling fraction might then be made larger in strata consisting of larger farms, since the variation between these would almost certainly be great; the statistical analysis would have to take account of different variances in different strata and, in the formation of final estimates, modified calculations appropriate to this variable sampling fraction would be necessary.

Strip sampling

In the sampling of row crops, and also in forestry, *strip sampling* is often convenient. The area to be sampled is regarded as divided into long narrow strips, which might be individual rows of a crop or arbitrary fixed widths of a forest. A number of strips is then selected at random, either unrestrictedly or in accordance with some stratification; for example, two rows might be selected at random out of the first 30, two out of rows 31 to 60, and so on. Sometimes the whole strip constitutes a sampling unit, sometimes only certain parts of it are used (as for example three 5-yard lengths of row).

Cost Functions

In the planning of any sampling investigation, the relative cost of alternative procedures, or the relative precision to be expected from alternative procedures having the same cost must be considered. The cost may be assessed in monetary terms, in time, or in any other convenient units. Since any desired precision could always be obtained by taking a large enough sample, the choice between alternative procedures must always take account of economic factors. An expression of the cost of sampling in terms of the constituent items of the design is known as a *cost function*. Examples have been given of a cost supposed proportional to the total area of the sample or related in some other way to the size and number of sampling units. When stratification is introduced, more elaborate cost functions may be needed, because the total cost may depend upon how many strata have to be defined and sampled as well as on the number, size, and shape of sampling units. Even a crude guess at the form of a cost function, however, is likely to be valuable in suggesting the most economical design.

Pilot Samples

Almost every new application of sampling produces special problems; the statistical principles remain the same but their application is affected by difficulties characteristic of the material to be sampled. If a statistician is consulted well in advance, he ought to be able to express a useful opinion on whether a proposed scheme of sampling is likely to give adequate answers to the questions at issue, and what modifications are desirable. Even for a totally new sampling problem, he may be able to make helpful suggestions. For a large investigation, however, full-scale work may be undesirable until preliminary trials have been examined in order to gain some idea of the magnitude of S.Ds. and some experience of difficulties and costs. In large surveys especially, a *pilot sample* of a small part of the population is often used to provide information that enables decisions on the finer points of design for the main sample to be taken wisely.

Systematic Sampling

In all fields of application of sampling techniques, we encounter publications that argue for systematic instead of random, selection of sampling units. For instance, in the problem of estimating the yield of a crop of wheat, discussed at length in this chapter, would not 35 square yard units spaced in a regular formation over the field in 5 lines of 7 units each at regular intervals be better than 35 units taken entirely at random? An arrangement of this type is shown in Figure 8.4, which may be contrasted with Figure 8.1. In the strip sampling of a row crop, would we not be wiser to take simply every 15th strip, instead of a random selection of 2 out of every 30 as was suggested earlier? Such a sample might consist of rows numbers 8, 23, 38, 53, and so on. Samples of this kind are termed *systematic samples*, and have been widely advocated for various reasons. They are often easier to use and are less subject to mistakes in the field, especially when the investigator is dependent upon assistants who are not very highly skilled. The trouble of making random selections is avoided. Most important of all, the sampling units are widely and evenly distributed over the population sampled, so that they may be expected to be thoroughly representative of all parts of it; therefore, say the advocates of systematic sampling, the estimate of mean yield (or whatever else is being studied) is likely to be more precise than that from a random sample of the same size.

Many writers on the subject have assumed that data from a systematic sample can be analysed in exactly the same way as if the units had been randomly selected, and they have based their supposed verifications of higher precision upon analyses of this kind. Unfortunately, this assumption is entirely without foundation and may sometimes be seriously misleading. *An element of randomization in the selection of sampling units is essential to the whole of the usual argument on which the calculation of standard deviations and standard errors is*

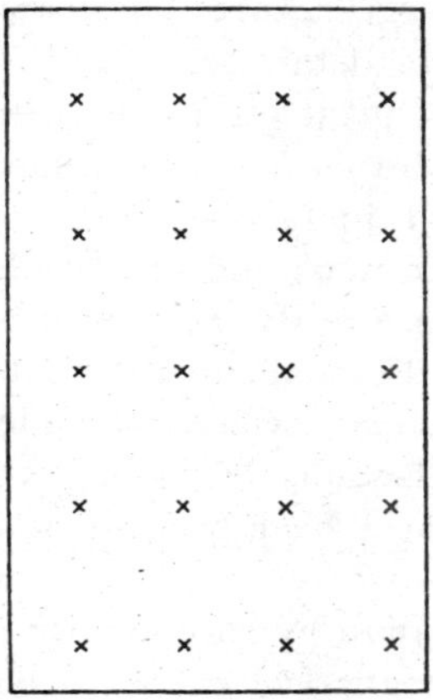

Fig. 8.4: Systematic sample.
Positions of 20 sampling units in a rectangular field.

based. The question is too complicated to be discussed adequately here, but fuller explanation has been given in my paper to which I referred earlier. Briefly we may state that assessment of the S.E. of an estimate from a systematic sample is impossible from the internal evidence of the sample alone, unless further assumptions on the nature of the variation between sampling units are made. These assumptions cannot easily be justified, and in any event are very little use unless the sample contains a large number of units.

The great merit of random sampling, whether stratified or not, is that it permits the precision of an estimate to be assessed, either as a standard error or as fiducial limits, by means of calculation from the data of one sample only. The investigator who adopts a systematic scheme for his sampling almost certainly sacrifices this possibility. Sometimes the sacrifice may be worth while, but he should not accept it lightly. He may have reason to believe that a systematic sample would give an estimate more precise than a random sample of the same size or of the same cost (though this is not invariably true), but he will often be wiser to have the random sample and know the precision than to have the systematic sample and not know anything about the precision. *Certainly he should never adopt a systematic sampling scheme without most careful consideration of its consequences, and after consultation with a statistician.*

Sample Surveys

Earlier in this chapter, I mentioned the use of sampling for extensive surveys in connexion with agriculture. These are so varied in type that I shall not discuss them in detail here; their difficulties are usually not so much of statistical principle as of practical detail. With the aid of a carefully planned design for the selection of a sample of farms, of fields, or of animals within farms, information on many different aspects of agricultural conditions and practice has been obtained. The survey of fertilizer practice already mentioned is an excellent example of what has been done in England. In the United States, estimates of crop yields over wide areas have been obtained by taking small samples of crops from random selections of fields; similar investigations of crop yields and crop acreages have been made in India and no doubt in other countries.

Pilot investigations, followed by careful planning for the main sampling scheme, enable the statistical errors in large surveys to be adequately controlled; the investigator may say in advance what degree of precision he expects to achieve on the items of chief interest, and, at the end of his survey, he may assign numerical measures of precision to every estimate he quotes.

The Sampling of Experiments

Even in the conduct of field experiments, sampling techniques are sometimes required. For example, if the experiment is a yield trial, circumstances may make complete harvesting of plots impossible; the experimenter may instead content himself with taking a sample of the produce of each plot, and comparing the sample estimates of yields instead of the complete plot values. The methods of yield estimation suggested earlier in this chapter can usually be adapted for this purpose, although the number of units per plot is likely to be substantially less than the number that would be taken for estimating the yield of a whole field. The precision of comparisons between the experimental treatments will be less than if whole plots had been harvested, because of the imperfect knowledge of the yield of each plot, but often the loss will not be large relative to the variation between plots. Sampling of plots for yield estimation, however, is scarcely to be recommended as a general practice, because it can be very laborious and may be little or no overall saving on the harvesting of complete plots. It sometimes provides the only way out of the difficulty that arises when unforeseen conditions make complete harvesting totally impossible.

Even when sample estimation of yields in an experiment is not contemplated, sampling may often be valuable for subsidiary measurements. An experimenter may wish to follow the growth of his crop throughout the season by studying such factors as germination, appearance of plants above ground, height of plants at various stages, dates of tillering, ear emergence, flowering, and so on; he may also wish to measure different components of yield such as (for beans) number of plants per acre, number of stems per plant, number of pods per stem, number of beans per pod, and weight per bean. Determination of quantities like these for whole plots will be quite impracticable, and measurements must be made on small sample areas or on samples of the plants.

In other circumstances, an experimenter may be in effect basing his comparison of treatments on samples without realizing it. In an experiment on the feeding of dairy cattle, the experimenter would naturally wish to compare butter-fat percentages for the various treatments; these would inevitably be measured on small samples of the milk produced. Sampling of plant material for comparison of chemical constitution is also very common in the detailed study of field experiments.

CHAPTER 9

CORRELATION AND REGRESSION

"Well then (quod Maister Moore) howe saye you in thys matter? what thincke ye to be the cause of these shelves and flattes that stoppe up *Sandwiche* haven? Forsoth syr (quoth he) I am an olde man, I thyncke that *Tenterton* steeple is the cause of *Goodwyn* sandes. For I am an old man syr (quod he) and I may remember the building of *Tenterton* steeple, and I may remember when there was no steeple at al there. And before that *Tenterton* steeple was in building, there was no maner of speaking of any flats or sands that stopped the haven, and therfore I thinke that *Tenterton* steple is the cause of the destroying and decaying of *Sandwych* haven."

Hugh Latimer: Last sermon preached before King Edward VI.

The discussion of statistical analyses of continuous variates in earlier chapters has been concerned solely with one variate at a time, — with butter-fat percentages or with yields of wheat or with weights of animals or with percentage disease infestations, but never with two variates simultaneously. In many investigations, however, the interrelationships of two or more variates are an important or even the essential object of study. Is there any relationship between yield and percentage butter-fat in a herd of cattle, in the sense that the high yielders on an average give milk of high fat content or alternatively that the high yielders tend to give milk of low fat content? Are the yields of grain and of straw in a series of wheat fields independent of one another? Do sugar-beet yields show any association with the numbers of plants per acre? Are the milk yields of cows in their first lactations associated with those of their dams? Does the yield of potatoes on a series of fields show any association with various physical and chemical analytical data for the soils or with the degree of virus infection of the crops?

To questions like these, answers are frequently required in agricultural research. Moreover, a simple affirmative or negative answer is rarely enough, for the investigator usually wants to obtain information on the closeness of any relationship and on its nature. The appropriate methods of statistical analysis are those of correlation and regression, and these will be briefly discussed in this final chapter.

Statistical Dependence

If two distinct measurements are recorded for each of a number of individuals (animals, fields, plots, plants, etc.), each will have its own probability distribution. If now we confine attention to individuals having a restricted range of values for one variate, we may find that the distribution of the other variate is different from its distribution for all individuals. The two variates are then said to exhibit *statistical dependence*. For example, records of total yield and mean butter-fat percentage might be obtained from a number of cows in their first lactations. The observed frequencies would correspond to the two probability distributions for these variates. If now the investigator restricts his attention to the high-yielding cows only (say to those whose yields exceed 900 gallons), he can study the distribution of butter-fat percentages amongst these; similarly he might study the butter-fat distribution amongst low-yielders or amongst a group with intermediate yields. If all the probability distributions so defined are identical (and therefore the same as the distribution for all cows), the two variates are statistically independent. Any departure from this independence would most commonly take the form of some type of *correlation* of the variates. Thus cows with high yields might give, on an average, lower fat percentages than those with low yields; note that the statement refers to averages and does not exclude the possibility that a few high-yielding animals may give high fat percentages. In other circumstances, a positive association might be found, rather than a negative, the high-yielders being now the animals with high butter-fat percentages; yet again, a more complicated relationship might appear, such as a tendency for fat percentages to be low for the very low yielding cows, higher for the moderate cows, but less again for the very high yielders. All these and other types of association are examples of correlated variates.

Statistical dependence and correlation are defined in terms of the probability distributions of the two variates. If records of yield and butter-fat are available for, say 150 cows, knowledge of the distribution is incomplete and only the observed distributions of frequencies can be studied. Consequently tests of significance of the evidence for correlation are required: the fact that three cows with yields of 700, 800, and 900 gallons gave 3.0 %, 3.2 %, and 3.6 % butter-fat could scarcely be regarded as strong evidence of a positive correlation between yield and butter-fat in the population from which these were selected. Some method of giving quantitative expression to the weight of evidence is essential for the interpretation of observational data.

Correlation and Causality

The reader must realize that statistical dependence of two variates is a way of describing properties of the two sets of numbers, and does not necessarily imply a causal dependence of one quantity measured upon the other. The discovery that, in a certain breed of cattle, butter-fat percentage is negatively correlated with yield (i. e., on an average, high-yielders are poor in butter-fat) does not in any sense mean that the production of a high yield 'causes' the fat content of the milk to decline. Illustrations of the fallacy in this argument are easily found. For instance, in the period 1924—1940, the number of notified mental defectives per million of population in England and Wales shows a close positive correlation with the number of wireless licences; only the most prejudiced, however, would argue that increased support of the British Broadcasting Corporation in that period came primarily from mental defectives or even that increased listening was responsible for the higher rate of mental deficiency! The correlation is produced because both quantities increased steadily during the period, the one possibly on account of refinements of diagnosis and complex social phenomena, the other because of the growth in popularity of a scientific invention newly available to the public. Many examples of such *nonsense correlations* can be found in sets of data showing increases or decreases with time. Another class of misleading correlations contains the *spurious correlations* produced when the two sets of values to be examined have been obtained as the results of division by a common value: this is particularly likely to occur in the analysis of indices formed from ratios of measurements, as for example if three measurements are made on each of a group of animals and the ratio of the first and third is then examined for correlation with the ratio of the second and third. Both nonsense and spurious correlations represent genuine numerical properties of the data, and their faults lie in the temptation to interpret them in terms of causal connexions of phenomena. If measurements of Goodwin sands at intervals during the building of Tenterton steeple existed, they might indeed show close correlation with the increasing height of the steeple, but few would contend that the opinion expressed in the quotation at the head of this chapter was thereby justified.

Nevertheless, the discovery of significant evidence for a correlation can be useful in directing attention to the existence of causal effects, either of the one quantity measured affecting the other or of both being subject to a common cause. In the milk example discussed, the nature of any correlation found might be a useful clue in the study of the physiological processes underlying milk production. Moreover, even though no simple causal connexion be likely, the existence of a correlation provides a basis for prediction and estimation, particularly if the association be close. If yield and butter-fat percentage are uncorrelated, knowledge that a cow is a high yielder is no help in guessing the butter-fat content of her milk, which is just as likely to be high, low, or average as that of any other cow. If a correlation has been found for other cows of the same population (e. g. the same breed or herd), knowledge of total yield for one cow gives some idea of how the fat content of her milk is likely to differ from average without any direct determination of the percentage. Here any correlation is unlikely to be close, so that the estimation would not be very satisfactory. More useful in practice might be a knowledge of the correlation (if any) between yields for the first and second lactations of the same cow. If a moderately close positive association were found, this could be used to predict the future yields of cows that had only just completed their first lactation, and so to give guidance on the culling of a herd. (In this instance, the same causes may be influencing both measurements, but that is quite immaterial from the point of view of the prediction). For this purpose, the algebraic form of the association must be estimated, most usefully by means of a regression equation. Once again the test of significance and the process of estimation go hand-in-hand: unless the evidence for correlation is significant, the existence of any true correlation may be in doubt and the prediction may be no better than if the information on first lactations were ignored, but when significance has been established the estimation process enables the data to be used to greatest practical advantage.

The Correlation Coefficient

One of the commonest ways of measuring the closeness of association between two variates is by means of their *correlation coefficient.* Table 9.1 contains data from ten plots of wheat on grain yields and the percentage protein for each plot. Inspection of the table, or, even more clearly, inspection of the diagrammatic representation in Fig. 9.1, shows that plots with high yields tend to have low protein contents. Although no simple rule of exact calculation is likely to subsist between these variates, the points in Fig. 9.1 lie roughly around a straight line and apparently no great violence would be done to the data if the true protein value for each plot were read from such a line instead of being directly determined. (The line shown in the diagram has been calculated in a manner that will be described later, but a similar line could be drawn by judgement of the eye alone).

The correlation coefficient is calculated by processes very like those already familiar for standard errors. Let us call the two variates in Table 9.1 x and y. The first four steps require no comment, since they are merely the formation of totals and sums of squares of deviations for each, but the reader should verify the arithmetic (as before, the symbol S indicates summation):

$$
\begin{aligned}
S(x) &= 14.3 + 12.8 + 12.7 + \ldots\ldots + 12.2 \\
&= 121.9, \\
S\left\{(x-\bar{x})^2\right\} &= (14.3)^2 + (12.8)^2 + (12.7)^2 + .. + (12.2)^2 - \frac{(121.9)^2}{10} \\
&= 1{,}513.61 - 1{,}485.96 \\
&= 27.65, \\
S(y) &= 127.3, \\
S\left\{(y-\bar{y})^2\right\} &= 1{,}648.05 - 1{,}620.53 \\
&= 27.52.
\end{aligned}
$$

Next a sum of products of corresponding deviations for x and y, that is to say of $(x-\bar{x})$ and $(y-\bar{y})$, is formed. By a rule exactly like that shown for the sum of squares at the end of Chapter 5,

$$
\begin{aligned}
S\left\{(x-\bar{x})(y-\bar{y})\right\} &= S(xy) - \frac{\{S(x)\}\{S(y)\}}{N} \\
&= 14.3 \times 10.8 + 12.8 \times 11.4 + \ldots + 12.2 \times 11.8 \\
&\quad - \frac{121.9 \times 127.3}{10} \\
&= 1{,}525.46 - 1{,}551.79 \\
&= -\ 26.33.
\end{aligned}
$$

Whereas sums of squares of deviations are always positive, a sum of products of deviations can be negative (and, indeed, must be when the association between the variates is of the negative character apparent in Fig. 9.1). The correlation coefficient, usually denoted by r, is defined by

Table 9.1: Yields and protein percentages for ten plots of wheat.

Plot no.	Yield (cwt./acre)	% protein
1	14.3	10.8
2	12.8	11.4
3	12.7	13.0
4	10.6	14.6
5	10.7	13.8
6	13.0	12.2
7	14.4	10.7
8	12.5	12.8
9	8.7	16.2
10	12.2	11.8

$$r = \frac{S\{(x-\bar{x})(y-\bar{y})\}}{\sqrt{S\{(x-\bar{x})^2\}\,S\{(y-\bar{y})^2\}}},$$

and so here its numerical value is

$$\begin{aligned} r &= \frac{-26.33}{\sqrt{(27.65 \times 27.52)}} \\ &= -\frac{26.33}{27.58} \\ &= -0.955. \end{aligned}$$

The definition of the correlation coefficient is such as to compel it to lie between the limits + 1 and — 1, whatever the number and nature of the observations. If the points in Fig. 9.1 lay absolutely on a straight line, the value of r would be — 1, whereas, if they lay perfectly on a line of positive slope (i. e. protein content increasing as yield increased; cf. Fig. 9.2), r would be + 1. A value of 0 corresponds to complete absence of linear association. From some points of view, r^2 is more readily interpretable than r as a measure of the closeness of association, since it is essentially the proportion of the total variability in one variate that can be accounted for by variations in the other. For Table 9.1

$$\begin{aligned} r^2 &= (-0.955)^2 \\ &= 0.912, \end{aligned}$$

so that 91 % of the variation in protein content can be explained as completely associated with variation in yield whereas 9 % is residual variation independent of yields.

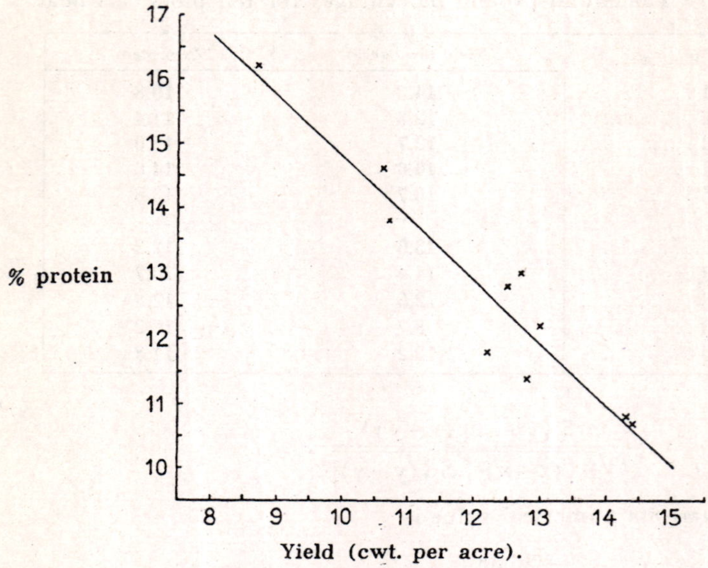

Fig. 9.1: Correlation diagram to show the data of Table 9.1.
The line represents the regression equation of protein percentage on yield.

Notice that the value of r is independent of the scales of measurement. If yields had been expressed as tons per acre or as grams per square metre, or if protein content had been expressed as parts per thousand, the totals and the sums of squares and products for x and y would have been different but r would still have been -0.955. The fact that $S\{(x-\bar{x})^2\}$ and $S\{(y-\bar{y})^2\}$ in this example are nearly equal is entirely coincidence, but $S\{(x-\bar{x})(y-\bar{y})\}$ must always be intermediate in magnitude between them.

As might be expected, a test of significance of r must depend upon the number of pairs of observations. A value as extreme as -0.9 could easily occur by chance if only 4 plots had been studied, but would almost certainly indicate some real association if it had been calculated from 100 plots. The correlation coefficient as calculated is an estimate of a true value for the whole population. If the null hypothesis is made that the true value is zero, so that the two variates have no linear association in the population, the significance of the evidence against the hypothesis may be tested by reference to standard tables of values of r corresponding to various levels of probability. Table 9.2 is an extract from the larger tables available. For a correlation coefficient, the number of degrees of freedom is 2 less than the number of pairs of observations, so that the value of r for Table 9.1 has 8 degrees of freedom. The significance levels in Table 9.2 are to be read as either positive or negative; with 8 d. f., any value greater than 0.63 or less than —0.63 is significant at a probability of 0.05, and the value —0.955 is therefore significant (in fact, it is well beyond the 0.01 level)*. Fiducial limits for the true correlation coefficient can be calculated from r by a method that will not be described; in the

Table 9.2: Some significance levels for r.

Degrees of freedom	No. of pairs	Probability 0.1	0.05	0.01
1	3	0.988	0.9969	0.9999
2	4	0.90	0.95	0.99
3	5	0.81	0.88	0.96
4	6	0.73	0.81	0.92
6	8	0.62	0.71	0.83
8	10	0.55	0.63	0.76
10	12	0.50	0.58	0.71
15	17	0.41	0.48	0.61
20	22	0.36	0.42	0.54
30	32	0.30	0.35	0.45
60	62	0.21	0.25	0.32

example, they are —0.800, —0.989, so that the true proportion of variation in protein content associated with variation in yield almost certainly lies between 64 % and 98 %.

As a second example, consider the data in Table 9.3, plant densities and yields of sugar on fully manured plots of sugar beet at 14 different centres in 1933. Fig. 9.2 shows these data diagrammatically, and suggests a positive association although this is by no means as clearly

* The test is most easily made by reference to a table like Table 9.2, but an exactly equivalent test can be performed by calculating

$$t = r\sqrt{\frac{n-2}{1-r^2}},$$

where n is the number of pairs of observations, and using Table 5.5 with (n — 2) d. f. Here

$$t = -9.1,$$

which is far beyond the 0.01 probability level for 8 d.f. The significance of a correlation coefficient should *not* be tested by use of the formula for the S.E. of r given in some books.

Table 9.3: **Plant densities and yields of sugar beet at fourteen centres in 1933.**

Centre no.	Plants (1,000 acre)	Sugar (cwt. acre)
1	23.4	35.3
2	27.2	44.6
3	30.8	49.3
4	25.9	54.1
5	26.3	52.8
6	22.6	41.7
7	22.8	26.9
8	26.3	54.4
9	22.8	51.9
10	37.8	71.9
11	29.3	54.9
12	23.8	54.4
13	30.1	47.6
14	34.4	49.5

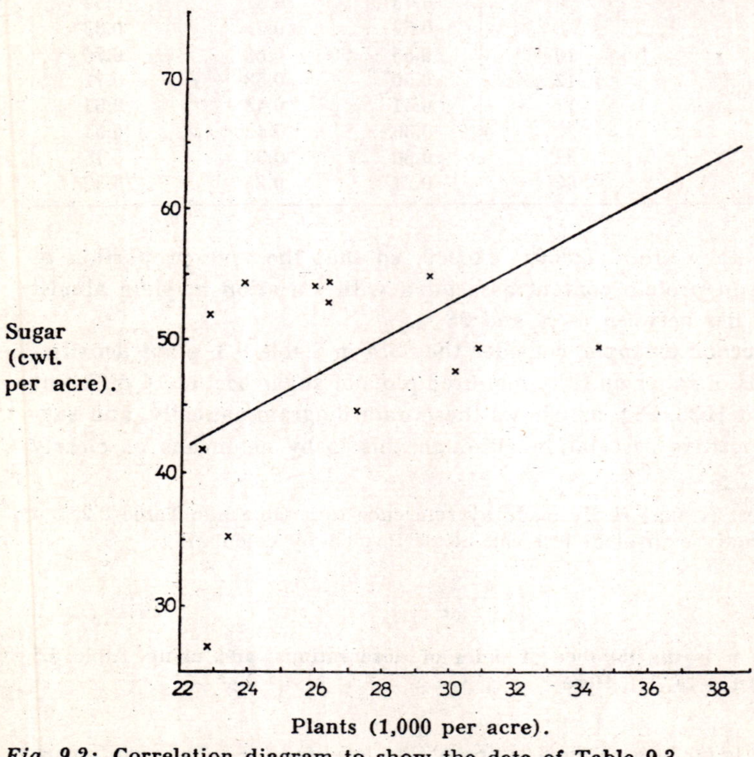

Fig. 9.2: Correlation diagram to show the data of Table 9.3. The line represents the regression equation of sugar yield on plant density.

marked as the correlation in Fig. 9.1. The line drawn in the diagram possibly acts as an additional persuasion to the eye, and without it the existence of any real correlation might be in some doubt. Calculations, which the reader should verify, give

$$S(x) = 383.5,$$
$$S\{(x-\bar{x})^2\} = 278.65,$$
$$S(y) = 689.3,$$
$$S\{(y-\bar{y})^2\} = 1{,}416.67,$$
$$S\{(x-\bar{x})(y-\bar{y})\} = 391.96.$$

(Note that the sum of products of deviations is now positive). Hence

$$r = \frac{391.96}{\sqrt{(278.65 \times 1{,}416.67}}$$
$$= 0.624.$$

The 0.05 significance level for 12 d. f. is 0.53, so that the correlation coefficient is significantly different from zero but by no means as markedly as in the other example (Here $t = 2.77$ with 12 d. f.). The fiducial limits are 0.119 and 0.862, so that variation in sugar yields may be estimated to be statistically dependent on variations in stand to the amount of 39 % ($0.624^2 = 0.389$), the true amount almost certainly being between 1 % and 74 % (scarcely a very precise estimate!).

A Cautionary Remark

The novice in statistics is apt to think that, because a correlation coefficient must lie between — 1 and + 1, a value as great as 0.5 (or — 0.5) is 'large', and may even speak of 0.4 or 0.3 as 'indicating close correlation'. As pointed out above, the square of a correlation coefficient gives a measure of the extent to which the two variates are tied to one another, so that a value of 0.5 means only that 25 % of the variation in one measurement is accounted for by variation in the other and a value of 0.3 means that less than 10 % is so accounted for. Correlation coefficients like these are often encountered, and may be shown as significant if the number of observations is large; they may indeed be important in elucidating the structure of causal and other relationships of measurements, but they scarcely indicate a degree of correlation close enough to be much use in prediction. For example, a discovery that the correlation coefficient between milk yields of cows in their first and second lactations was 0.35 could be a perfectly sound statement of fact yet would not be much help in predicting yields for the second lactations of cows which had only just completed their first.

I suggest that, in so far as exact sense can be given to phrases of general description, the words 'close correlation' should not be used in relation to any correlation coefficient smaller than 0.7; (this represents about 50 % dependence of variation, and is about the value to be expected for a correlation between parent and offspring in animals or plants for a character largely determined by heredity and but little by environment. Coefficients greater than 0.85 might be described as referring to 'very close correlation', since 70—75 % or more of the variations in the two measurements are then tied together. Coefficients as large as this are seldom found, however, except for pairs of measurements having some obviously important physical connexion.

Curvilinear Correlation

If the points in a correlation diagram like Figs. 9.1 and 9.2 were to suggest a curved relationship rather than a straight line, the correlation coefficient might be completely inappropriate as a measure of association. Large deviations from a straight line do not matter as long as these are not of such a systematic character as to indicate a curve. Data such as those in Table 9.4, however, when exhibited diagrammatically in Fig. 9.3, show a type of relationship entirely different from that for Table 9.1 (Fig. 9.1). There is again undoubtedly a close correlation or association between the yields and the protein contents, but the correlation coefficient is quite inappropriate to the expression of this; in fact, $r = -0.06$, which without the diagram might suggest a negligible degree of association.

In correlation studies, a diagram should always be drawn, unless the investigator is sufficiently experienced to visualize it from inspection of a table. Even to the investigator who does not intend to make any statistical calculations himself, a diagram may be very illumi-

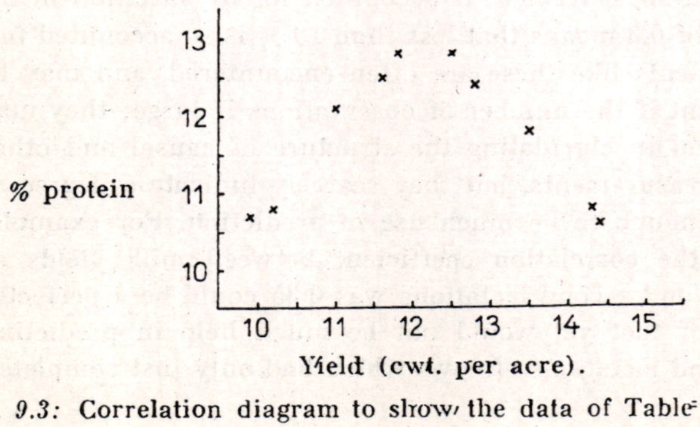

Fig. 9.3: Correlation diagram to show the data of Table 9.4.

nating, although without the appropriate calculations and tests he may be tempted to see correlations where none exist. Any serious indications of non-linearity should be submitted to a statistician before correlation coefficients are computed.

Table 9.4: Second series of yields and protein percentages for ten plots of wheat.

Plot no.	Yield (cwt./acre)	% protein
1	14.3	10.8
2	11.6	12.5
3	12.5	12.8
4	13.5	11.8
5	9.9	10.7
6	10.2	10.8
7	12.8	12.4
8	11.8	12.8
9	11.0	12.1
10	14.4	10.6

Correlation Tables

When the number of individuals measured is large, detailed presentation of the data in the same form as Tables 9.1 and 9.3 would take a great amount of space. Instead, a *correlation table* is often used, this being an obvious generalization of the grouped frequency table in Table 3.3. Examples are given in Tables 9.5 and 9.6. Such tables are constructed by grouping all individuals into short ranges of one variate, and then counting how many fall into short ranges of the other. Thus Table 9.5 refers to 586 Ayrshire cows which were first classified according to age; 16 were between 1½ and 2½ years, 124 between 2½ and 3½ years, and so on. Of those in the 2½—3½ group, 2 had a weekly average yield between 6 and 8 gallons, 8 gave between 8 and 10 gallons, 19 gave between 10 and 12 gallons, and so on. As for Table 3.3, finer gradations of scale could be adopted, but these would increase the size of the table and give many more empty spaces, whereas coarser gradations would lose much information.

Inspection of a correlation table is often helpful as an indication of the closeness and form of an association. Thus, both Table 9.5 and Table 9.6 obviously indicate positive correlations (for a negative correlation, the bulk of the entries would lie near the other diagonal of the table), that for Table 9.6 being much closer than that for Table 9.5. Both relationships are approximately linear, though a test of significance would be needed in order to examine this point thoroughly. The

Table 9.5: Correlation table for age and average weekly milk yield of 586 Ayrshire cows.
Entries in the table are numbers of cows in the various groups; absence of an entry is to be read as zero.

Weekly yield (gallons)	Age (years)																Total
	1½ –2½	2½ –3½	3½ –4½	4½ –5½	5½ –6½	6½ –7½	7½ –8½	8½ –9½	9½ –10½	10½ –11½	11½ –12½	12½ –13½	13½ –14½	14½ –15½	15½ –16½	16½ –17½	
6— 8	1	2	3														6
8—10		8	2	3										1			14
10—12	1	19	12	5	3												40
12—14	3	29	24	12	4	3		2	2					1			80
14—16	7	38	23	26	10	4	3	4	2	4		2	1				124
16—18	4	16	28	23	11	9	9	5	4	1	7		2				119
18—20		9	12	15	11	6	7	8	5	2	1	1		1			78
20—22		2	6	11	9	3	8	8	7	4	1	2		1			62
22—24		1	4	3	8	4	7	3	2	3					1	1	37
24—26			2		4	5	1	2		1							15
26—28				1	1	1	1		1		1						6
28—30					1		1										2
30—32											2						2
32—34																	0
34—36						1											1
Total	16	124	116	99	62	36	37	32	23	15	12	5	3	4	1	1	586

Table 9.6: Correlation table for length and breadth of 9,440 beans.
Entries in the table are numbers of beans in the various groups; absence of an entry is to be read as zero.

Breadth (mm.)	Length (mm.)																Total
	9.25 —9.75	9.75 —10.25	10.25 —10.75	10.75 —11.25	11.25 —11.75	11.75 —12.25	12.25 —12.75	12.75 —13.25	13.25 —13.75	13.75 —14.25	14.25 —14.75	14.75 —15.25	15.25 —15.75	15.75 —16.25	16.25 —16.75	16.75 —17.25	
6.25—6.50 ..	1	1	1				1										4
6.50—6.75 ..		3	4	1		2											10
6.75—7.00 ..		1	7	13	12	21	8	9	1								72
7.00—7.25 ..		1	6	11	32	25	35	28	18	13	1						170
7.25—7.50 ..		1		11	22	37	78	124	137	91	23	6					530
7.50—7.75 ..					4	27	55	175	361	469	236	65	4	1			1,397
7.75—8.00 ..						3	19	89	330	794	871	385	81	7			2,579
8.00—8.25 ..							3	12	73	362	913	956	375	44	4		2,742
8.25—8.50 ..									9	56	227	574	494	105	18		1,483
8.50—8.75 ..										2	23	93	156	101	23	2	400
8.75—9.00 ..													19	17	8	4	48
9.00—9.25 ..												3			2		5
Total	1	7	18	36	70	115	199	437	929	1,787	2,294	2,082	1,129	275	55	6	9,440

correlation coefficient can be evaluated by assuming that all observations are at the mid-points of their groups; thus, in Table 9.5, the third entry in the second column is used in calculation as though it referred to 19 cows each aged 3 years and yielding 11 gallons of milk per week. The correlation coefficients are 0.43 for Table 9.5 and 0.75 for Table 9.6, both being highly significant; these are calculated just as for the data of Table 9.1, although the arithmetic is more laborious.

Linear Regression

Correlation has been discussed before regression because so many people with only a slight knowledge of statistics have heard of (and may even know how to calculate) a correlation coefficient. In modern statistical science, however, the correlation coefficient is used much less than was once customary. Indeed, I have known more than one statistician to say that he never calculates correlation coefficients. Without adopting so extreme an attitude, we must nevertheless admit that regression analysis of data on the interrelationship of two variates is often of more practical use and more directly relevant to the purpose of an inquiry than is correlation anlysis. Sometimes both are wanted in order to examine different aspects of a problem, and again sometimes correlation is the more useful.

The correlation coefficient is a measure only of the closeness of a linear correlation between two variates, and tells nothing about the form of that relationship. This is illustrated by comparison of Table 9.7 and Fig. 9.4 with Table 9.1 and Fig. 9.1. Both sets of data relate to the protein contents of wheat grown on ten different plots, and both suggest a roughly linear relationship governing the decrease in protein content with increase of yield. Fig. 9.4, however, shows a much less extreme decrease than does Fig. 9.1; for example, in Fig. 9.1, the protein content decreases from about 16 % for yields of 9 cwt. per acre to about 11 % for yields of about 14 cwt. per acre, whereas in Fig. 9.4 the protein contents at these yields are about 13½ % and about 12¼ %. Nevertheless, the correlation coefficient for Table 9.7 is the same as that for Table 9.1! This is not an indication of a flaw in the definition of the coefficient but only a reflexion of its purpose. In fact, I constructed the 'data' in Table 9.7 from those in Table 9.1 by a simple device; I left the yields unaltered, but reduced the deviations of each protein percentage from the mean of the column to one-quarter of its former value, so that, for example, $(12.2 - 12.7) = \frac{1}{4}(10.8 - 12.7)$. If Fig. 9.4 were re-drawn with a fourfold magnification of the vertical scale, the configuration of points would appear exactly like that of

Fig. 9.1, and therefore the diagrams are rightly judged to have the same degree of correlation.

As soon as interest is focussed on the form of the relationship, the question of finding the best straight line to represent the statistical dependence of one variate on another arises. For the type of problem

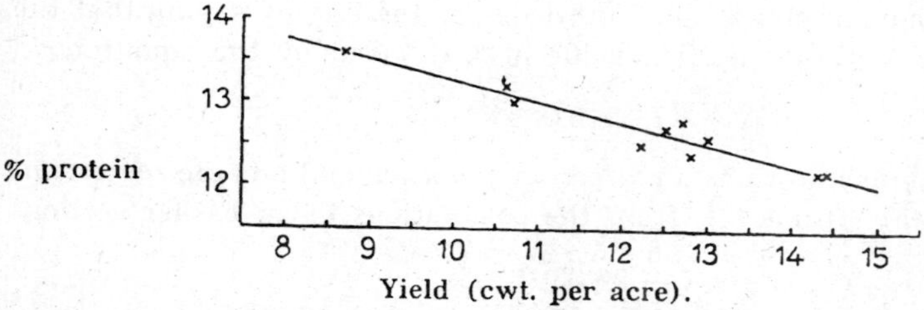

Fig. 9.4: Correlation diagram to show the data of Table 9.7. The line represents the regression equation of protein percentage on yield.

Table 9.7: Third series of yields and protein percentages for ten plots of wheat.

Plot no.	Yield (cwt./acre)	% protein
1	14.3	12.2
2	12.8	12.4
3	12.7	12.8
4	10.6	13.2
5	10.7	13.0
6	13.0	12.6
7	14.4	12.2
8	12.5	12.7
9	8.7	13.6
10	12.2	12.5

that we are considering, the answer is very simply obtained from calculations already made. Mathematical theory shows that, at a value of x equal to the mean for all the data, (i. e. at $\bar{x}$) the value of y for the line is the mean of this variate for all the data (i. e. $\bar{y}$); hence the line passes through the point ($\bar{x}$. $\bar{y}$). Secondly, if we define a quantity b as the ratio of two of the expressions used in calculating the correlation coefficient:

$$b = \frac{S\{(x-\bar{x})(y-\bar{y})\}}{S\{(x-\bar{x})^2\}},$$

then the theory shows that the value of y for the line increases by b for each unit increase in x. Thus the line is completely specified, since we know a point on it and its direction or slope. The line represents the statistical expectation or the expected value of y corresponding to every x (i. e. the average value of y to be expected for repeated observations having the same x); we usually write Y for this expectation, and can then summarize our knowledge of the line by saying that the point on it for any specified value of x is given by the equation

$$Y = \bar{y} + b(x - \bar{x}) .$$

The equation is known as a *regression equation,* and b is the *regression coefficient.* For Table 9.1, from the calculations in an earlier section,

$$b = \frac{-26.33}{27.65}$$
$$= -0.952,$$

(which here, quite by accident, is almost equal to the correlation coefficient), and therefore

$$Y = 12.73 - 0.952\,(x - 12.19),$$

which may be simplified to read

$$Y = 24.3 - 0.95\,x.$$

If values of x and Y satisfying this equation are plotted on Fig. 9.1, they will lie on a straight line, which line has been drawn there. For Table 9.7 and Fig. 9.4, the corresponding results are

$$b = \frac{-6.76}{27.65}$$
$$= -0.244$$

and

$$Y = 12.72 - 0.244\,(x - 12.19)$$

or

$$Y = 15.7 - 0.24\,x.$$

The values of $\bar{x}$ and $\bar{y}$ are the same for both sets of data, but the second regression coefficient is about one-quarter of the first (failing to be exactly one-quarter only because of the recording of y to the nearest 0.1 in Table 9.7).

Whereas the correlation coefficient is independent of the units in which the two variates are measured, a regression coefficient measures a change in y per unit change in x, and its unit of measurement is therefore "y-units per x-unit". For Table 9.1, the regression coefficient of protein content on yield is —0.95 % per cwt.-per-acre.

As soon as any prediction is required from data on two variates, the regression equation is needed. For example, if an investigator wishes to estimate what the protein content would be for wheat grown on a plot similar to those summarized in Table 9.1 and yielding 11.5 cwt. per acre, he will calculate

$$Y = 24.3 - 0.95 \times 11.5,$$

and use the result, 13.4 %, as his estimate. He may obtain exactly the same result by direct reading from the straight line in Fig. 9.1. This is an estimate of the average protein content to be expected from a large number of plots yielding 11.5 cwt. per acre, under similar conditions of experiment, and is subject to error in individual instances. Further statistical calculations not described here can be used to give a standard error for the estimate.

A linear regression equation is here evidently a reasonable expression of the statistical dependence of y on x, at least within the range of the observations. If the data were more numerous, as in Table 9.5, a test of significance of deviations from linearity could be applied, but here the data are too few and we must be content with the visual impression from Fig. 9.1. Nevertheless, we must not assume that linearity is an adequate rule outside the range of observations; for example, we should be most unwise to argue that a plot yielding substantially better than any of these indifferent plots and giving, say, 20 cwt. per acre would be expected to have only 5.3 % protein (as predicted by extending the regression line). Extrapolation of this kind not only leads to nonsense if carried far (thus for 30 cwt. per acre the value of Y is negative!), but it may give something very far from the truth long before the stage of nonsense is reached.

For Table 9.3,

$$b = \frac{391.96}{278.65}$$

$$= 1.407$$

and

$$Y = 49.24 + 1.407\,(x - 27.39)$$

or

$$Y = 10.7 + 1.41\,x.$$

This is a regression equation, from which may be estimated the yields of sugar on similarly manured plots at specified plant densities. The yield is estimated to increase by 1.41 cwt. per acre for each additional 1,000 plants per acre; of course, this is valid only over a restricted range of stands (in particular, over the range in Table 9.2 from about 23,000 to about 37,000 per acre), and would certainly break down for much denser stands where further increase in density might even reduce the yield.

†*The Two Regressions*

A correlation coefficient is a symmetrical statement about the interdependence of two variates; a regression coefficient is an asymmetrical expression of the statistical dependence of one variate on another. If, for the data of Table 9.2, the investigator wished to estimate the stand corresponding to a particular yield, he might want to use a different regression relationship, namely that of x on y. This is found in exactly the same way, but interchanging the roles of the two variates. It is

$$X = 27.39 + \frac{391.96}{1{,}416.67}(y - 49.24)$$

or

$$X = 13.8 + 0.28\,y.$$

This would be appropriate if he wanted to estimate, say, the mean density of plants on plots yielding 40 cwt. per acre, the answer being 25,000 per acre. On the other hand, if he wished to estimate what density of plants would on an average yield 40 cwt. per acre he would find the value of x that makes $Y = 40$ in the earlier equation and so would obtain the estimate 20,800 per acre. Careful thought is needed to distinguish these two ideas, and complications concerned with the choice of an appropriate regression usually need to be discussed with a statistician.

Independent and Dependent Variates

In regression analysis, the variate predicted or estimated (y in most of this chapter, but x in the last equation) is usually termed the *dependent variate,* and the variate on which prediction or estimation is based (x in most of this chapter, but y in the last equation) is termed the *independent variate.* Again only statistical dependence is implied, and no notion of cause and effect need enter. Often the user of regression methods has in mind the possibility of underlying causes, as when he studies the regression of weight-gains in pigs on quantities of food consumed, and then only the one regression equation is of interest. Often the notion of common causes or of inexplicable association is a truer description of his ideas, as in the association of yield and stand just discussed, the decision as to which variate is 'dependent' being then perhaps determined by the immediate purpose. Sometimes even a regression apparently opposed to the direction of cause and effect may be wanted for estimation, as in a regression equation of heights of fathers upon heights of their sons.

Additional difficulties arise if the variate that is to be regarded as independent is subject to errors of measurement. For example, if the sugar beet stands in the example above were based on counts of plants for the whole of each experimental plot, the regression analysis of yield on stand as given is correct. If instead only sample areas on each plot were counted, the values of x are no longer the true plant densities corresponding to the recorded yields but are only sampling estimates of these and, if the samples were small, may be subject to large sampling errors. The usual effect of this is to reduce the regression coefficient, so that b as calculated is likely to be an under-estimate of the true regression coefficient of yield on stand.

†*Standard Errors of Regression Coefficients*

The standard error of a regression coefficient may be found as follows. First a *residual variance* is estimated, this being a variance of values of y based upon their deviations from the regression line instead of upon their deviations from zero; it is

$$s^2 = \left[S\left\{(y-\bar{y})^2\right\} - \frac{\left[S\left\{(x-\bar{x})(y-y)\right\}\right]^2}{S\left\{(x-\bar{x})^2\right\}} \right] \div (n-2)$$

with $(n-2)$ d. f. The variance of b is then calculated as

$$\text{Variance of } b = \frac{s^2}{S\left\{(x-\bar{x})^2\right\}},$$

and the S.E. of b is the square root of this. The S.E. so found may be used in testing the significance of b, or in assigning fiducial limits, by the use of t with $(n-2)$ d. f. in the usual manner. For example, for Table 9.1,

$$s^2 = \left[27.52 - \frac{(26.33)^2}{27.65} \right] \div 8$$

$$= 0.306.$$

Hence

$$\text{S. E. of } b = \sqrt{\frac{0.306}{27.65}}$$

$$= 0.105.$$

Since b is about 9 times its S.E., the regression coefficient is certainly significantly different from zero. The 0.05 value of t for 8 d. f. is 2.31 (Table 5.5), and

$$2.31 \times 0.105 = 0.24.$$

Hence fiducial limits to b are 0.71, 1.19. The test of significance here is mathematically identical with that for the correlation coefficient given earlier in the chapter, the ratio of b to its S.E. being exactly the t-value explained in a footnote. If only a test of significance is needed, the easiest way of making it may be by use of the correlation coefficient and a table such as Table 9.2; if fiducial limits are wanted, these additional calculations of a standard error for b are necessary.

†*Curvilinear Regression*

The method of estimating a regression equation and discussing its properties that has been described in the previous section is applicable only when the relationship is approximately that of a straight line. Often a straight line is an adequate approximation over a fairly narrow range of observations, even though we know that it cannot be of wider validity. As already mentioned, statistical tests for the significance of deviations from linearity can be applied.

For data such as those of Table 9.4, illustrated in Fig. 9.3, linear regression is obviously unsuitable. More elaborate methods, such as that based upon *orthogonal polynomials,* can be adopted for finding a curvilinear regression equation. Thus, although no simple and satisfactory measure of the closeness of correlation, analogous to the correlation coefficient, is available for non-linear relationships, regression techniques can be extended to these and can be interpreted in much the same way as can linear regression. Even Fig. 9.2 suggests that a linear regression is not entirely satisfactory for the sugar beet data; except for one very high yield, the points indicate that the increase in yield for an additional 1,000 plants may be reduced at the higher plant densities, as would be expected from considerations of common sense!

†*Multivariate Analysis*

Occasionally the interrelationships of three of more variates must be studied simultaneously; for example, the yield of sugar from various plots of beet might be examined in relation to stand and to various chemical characteristics of the soil, or the yields of milk in a herd of cattle might be examined in relation to age of cow, duration of lactation, and percentage of butter-fat. The closeness of relationship and degree of association between different variates can be expressed by *multiple* and *partial correlation coefficients* and the nature of the relationship by a *multiple regression equation* with *partial regression coefficients* for each independent variate. To the statistician, these are obvious extensions of the simpler ideas already discussed, but the calculations are usually much more laborious.

†*Analysis of Covariance*

A very close connexion subsists between regression analysis and the technique of analysis of variance described in earlier chapters. One important consequence is that regression methods can be used to improve the precision of comparisons between treatments in an experiment. The procedure, the *analysis of covariance,* has already been mentioned in Chapter 6.

Suppose that in an experiment on animal nutrition the sizes of the animals available for experimentation vary quite widely. The investigator, in cooperation with a statistician, may design his experiment most carefully, taking advantage of litters or any other natural groupings of the animals in order to define blocks and randomizing properly within the constraints of his design. At the end of his experiment, however, he may suspect that certain treatments have done well, as shown by the measurement on which he wishes to base his conclusions (final weight, milk yield or quality, egg production, some measure of meat production, etc.), not so much because of inherent merit as because they were 'lucky' in being assigned to superior animals. The experiment is fair and unbiased, in the sense that one treatment was just as likely to be so favoured as any other. Nevertheless, if some animals were superior then one or more treatments would benefit from being tested on them; knowledge that in a long series of experiments all treatments would be about equally lucky is small comfort to the experimenter who can do only one!

If he has taken the precaution of recording the initial weight of each animal, he can estimate the regression coefficient of his final measurement on initial weight. This should not be calculated crudely from total sums of squares and products but from the 'error' line of an analysis of covariance table. The initial weights can be subjected to an analysis of variance *exactly* like that for the final measurements; no significant effects of treatments should appear, since the treatments had not been applied before the first weighing, but the same pattern of analysis can be followed. The sum of products of deviations for the two variates can similarly be analysed into components. Instead of an analysis of variance on y, the final measurement, an analysis of covariance with columns for the analysis of x^2, xy, and y^2 will appear, and the regression coefficient required is the ratio of the error sum of products to the error sum of squares for x, the initial measurement. This regression coefficient is an estimate of the effect of an increase of 1 unit in x on the value of y in the absence of treatment. By means of it, mean values of y for the various treatments can be adjusted so as to estimate what they would have been had every animal been alike in initial weight. If the data show a significant regression, the result of the adjustment will usually be to give smaller standard errors for differences between treatments and so to improve the precision of the experiment. As described here, the method may sound complicated and difficult, but in fact it is easily reduced to a simple routine within the framework of the analysis of variance.

The same technique may be applied in many other circumstances. For example, an experiment on sugar beet may chance to have great differences in stand from plot to plot for reasons having no connexion with treatments that have been applied: this might happen as a result of pest or weather damage that was restricted to a part of the experimental area. An analysis of yields without allowance for irregularities of stand would show large standard errors, while to adjust all yields on a basis of proportionality to stand would be unfair because individual roots can grow larger on the less densely populated plots. Adjustment to equal stands by means of a regression coefficient of yield on stand, calculated from an analysis of covariance, is likely to be the ideal way of rescuing as much information as possible from this unfortunate experiment, and of making comparisons between treatments with improved precision. In such an experiment, however, we need to make sure that the stand has not itself been appreciably affected by treatments; if it has, the regression adjustment would be eliminating part of the treatment differentiation as well as making allowance for the misfortunes that the experiment had suffered.

VALEDICTION

> "The student must be careful in calculations involving the decimal point to put it in its exact place, neither too much to the right nor too much to the left."
>
> *H. Belloc:* Caliban's Guide to Letters.

When you have understood the main features of the nine chapters in this book, you should have a reasonable appreciation of the function of statistical science and of the statistician in agricultural research. Even if your own abilities in statistical analysis are closely circumscribed, with a little practice you should be able to perform many of the basic types of statistical calculation. In closing, I ask you to remember the following twelve rules and to apply them in all quantitative investigations.

(1) When you propose to undertake an experiment or a sampling investigation of a kind that presents any novelty to you, consult a statistician at an early stage of your planning.

(2) At this early stage of planning, ask yourself why you are making the investigation and what kind of information you expect to obtain; nevertheless, keep your mind open to the possibility of obtaining other useful information, without prejudice to your main interests, by suitable choice of design.

(3) Whatever the nature of your investigation, choose for it the design that, in the light of existing knowledge, seems likely to give you the most useful and precise information on the subjects of inquiry. A ready-made design taken from the published account of a superficially similar investigation is likely to be far less satisfactory than one carefully chosen or constructed in relation to your own resources and objectives.

(4) When the arrangement of a design or the selection of a sample calls for randomization, adopt a strictly random scheme and not merely blind or haphazard selection; when in doubt whether randomization is necessary, randomize.

(5) Keep full records of all matters relating to the design and conduct of your investigation, and inform your statistician fully on these.

(6) Beware of the many opportunities for mistakes in copying from field note-books to permanent record files or to sheets for computation. Every new copy must be carefully checked against the original. I have known great trouble caused in statistical analysis by a neatly typed 'copy' of data that was unfortunately full of mistakes.

(7) Do not reject data merely because they seem extraordinary. You should never reject data without independent evidence that they are untrustworthy or unsuitable for inclusion with the remainder of the experiment. Generally rejection should be made only after discussion with your statistician, and always he should be told of any loss or discarding of records.

(8) If you do any of your own computing – as you certainly ought in order to understand your own experiments properly — make sure that all arithmetic is carefully checked. A mistake in arithmetic can lead to errors of interpretation of an experiment as serious as an error in the conduct of the experiment, and the one is no more excusable than the other.

(9) When you are experienced enough to make your own statistical analyses, be sure that you choose the right technique and not merely any one that you can remember! This may sound obvious, but not infrequently an entirely inappropriate method of analysis is applied (such as a χ^2 test on a continuous variate) apparently for no reason other than that the table of data was superficially of the same form as one in a text-book.

(10) In summarizing the results of a statistical analysis, use simple tables and diagrams; to put too many entries into one table or too many curves on one diagram ensures that few will read and fewer understand it.

(11) Express your conclusions in terms of the subject under investigation and its behaviour, with a minimum of statistical jargon; do not regard statements about mean squares, values of χ^2, and significant variance ratios as ultimate conclusions.

(12) Remember that no amount of statistical manipulation can make a bad experiment give good results. The first duty of the experimenter is to perform good experiments: he should not let a newly developed enthusiasm for statistics blind him to this. In this book, I have tried to show that the best results will be achieved by a combination of good experimentation and good statistical science, but I have no hesitation in asserting that the first is by far the more important.

REFERENCES

In order to avoid confusing the text of an elementary book with footnotes, and also to discourage any idea that the numerical examples were chosen for the importance of agricultural principles contained in them, no references to the sources of data and examples have been given earlier. These sources are listed below; if the reader looks up any one of the references, he should remember that the sole reason for its use in this book is that it provided an illustration of statistical technique or practice appropriate to the context. All data for which no reference is quoted below are entirely fictitious.

Table 2.6: Wireworms and Food Production, Bulletin no. 128, Ministry of Agriculture & Fisheries (1944).

Tables 2.9, 5.5, 6.3, 9.2: *R. A. Fisher & F. Yates,* Statistical Tables for Biological, Agricultural & Medical Research (1948. Oliver & Boyd, Edinburgh) and many other collections of statistical tables.

Table 3.1: *J. E. Kerrich,* An Experimental Introduction to the Theory of Probability (1946: Munksgaard, Copenhagen).

Table 3.2: My own experiment.

Table 3.3: *J. Wilson,* Scientific Proceedings of the Royal Dublin Society, vol. 12, pp. 470—479 (1910).

Fig. 6.3: Rothamsted Experimental Station Report for 1932, p. 160.

Table 6.8: *W. G. Cochran, K. M. Autrey, & C. Y. Cannon,* Journal of Dairy Science, vol. 24, pp. 937—951.

Fig. 7.1: Rothamsted Experimental Station Report for 1930, p. 148.

Fig. 7.2: » » » » » 1932, p. 180.

Fig. 7.3: » » » » » 1937, p. 162, (simplified).

Fig. 7.4: » » » » » 1933, p. 133.

Fig. 7.5: » » » » » 1935, p. 185.

Fig. 7.6: » » » » » 1930, p. 162.

Fig. 7.7: » » » » » 1932, p. 152.

Fig. 7.8: » » » » » 1932, p. 178.

Fig. 7.9: » » » » » 1935, p. 192.

Fig. 7.10: » » » » » 1934, p. 193.

Pig Experiment on p. 126: Rothamsted Experimental Station Report for 1935, p. 206.

Table 9.1 & Fig. 9.1: Adapted from *A. J. de Oliveira,* Agronomia Lusitana, vol. 8, pp. 147—159 (1946).

Table 9.3 & Fig. 9.2: Rothamsted Experimental Station Report for 1934, pp. 218—221.

Table 9.5: *H. D. Vigor,* Journal of the Board of Agriculture, Supplement no. 11 (1913).

Table 9.6: *S. J. Pretorius,* Biometrika, vol. 22, pp. 109—223.

INDEX

This book is intended for consecutive reading and not as a work of reference. The index therefore lists only the pages on which the more important discussions of the various subjects are to be found. Entries for subjects that appear wholly agricultural (e. g. particular crops) relate only to discussions of actual problems or data and not to every mention of these that is made by way of illustration.